GLOBALIZATION, LOCALIZATION AND SUSTAINABLE LIVELIHOODS

Globalization, Localization and Sustainable Livelihoods

Edited by
REIDAR ALMÅS
Norwegian University of Science and Technology, Norway
GEOFFREY LAWRENCE
The University of Queensland, Australia

ASHGATE

Published by
Ashgate Publishing Limited
Gower House
Croft Road
Aldershot
Hants GU11 3HR
England

Ashgate Publishing Company
Suite 420, 101 Cherry Street
Burlington, VT 05401-4405 USA

Ashgate website: http://www.ashgate.com

British Library Cataloguing in Publication Data
Globalization, localization and sustainable livelihoods
1.Sociology, Rural 2.Globalization 3.Sustainable development 4.Rural development
I.Almås, Reidar II.Lawrence, Geoffrey, 1950- III.World Congress of Rural Sociology (10th : 2000 : Rio de Janeiro, Brazil)
307.7'2

Library of Congress Cataloging-in-Publication Data
Globalization, localization, and sustainable livelihoods / edited by Reidar Almås and Geoffrey Lawrence.
     p.cm.
   Includes bibliographical references and index.
   ISBN 0-7546-0948-0 (alk. paper)
     1. Rural development--Sociological aspects. 2. Sustainable agriculture. 3. Globalization. I.Almås, Reidar, 1943- II. Lawrence, Geoffrey.

HN49.C6 G57 2002
307.1'4--dc21
ISBN 0 7546 0948 0
2002074545

Printed and bound by Athenaeum Press, Ltd., Gateshead, Tyne & Wear.

Contents

Figures and Tables   vii
Notes on Contributors   viii
Acknowledgements   xiii

PART I: GLOBALIZATION AND FOOD

1   Introduction: The Global/Local Problematic
    Reidar Almås and Geoffrey Lawrence   3

2   Fast Food/Slow Food: Standardizing and Differentiating
    Cultures of Food
    Mara Miele and Jonathan Murdoch   25

3   Restructuring of the Sugar Supply to the Industrially-
    Processed Food Chain: The Brazilian Case
    Maria Rita Pontes Assumpção   43

4   Forging New Local/Global Links Through Fair Trade
    Agro-Food Networks
    Laura T. Raynolds   57

5   The Power of Food
    Philip McMichael   69

PART II: THE RESTRUCTURING OF LOCAL AGRICULTURE

6   Assessing the Neoliberal Experiment in Antipodean
    Agriculture
    Hugh Campbell and Geoffrey Lawrence   89

7   Resisting Integration in the Global Agro-Food System:
    Corporate Chickens and Community Controversy in Texas
    Douglas H. Constance, Alessandro Bonanno,
    Caron Cates, Daniel L. Argo and Mirenda Harris   103

8   Globalization and Agro-Food Systems in Argentina
      Miguel Teubal and Javier Rodríguez   119

9   Technological Transformation of Household Milk
      Production in the South of Brazil: Social Impacts
      Ivaldo Gehlen   135

10   Co-operatives to Companies: The South African Wine
      Industry in the Face of Globalization
      Joachim Ewert   153

**PART III: COMMUNITIES AND RESISTANCE IN A
         GLOBALIZING WORLD**

11   Two Roads to the Global Village: A Comparison of How
      a Coastal and a Mountain Region of Norway Have Found
      Strategies to Cope with Globalization
      Reidar Almås   173

12   Local Consequences and Responses to Global Integration:
      The Role of the State in the Less Favoured Zones
      Manuel Belo Moreira   189

13   Global Change, Social Exclusion and Sustainable
      Livelihoods in Rural Britain
      Mark Shucksmith   205

14   Countering Localized Impacts of Globalization: Some
      Rural Community Development Initiatives in Australia
      Erica Hallebone, Mary Mahoney and Mardie Townsend   221

15   The Problem of Trust at the Margins of the Network
      Economy: Trust and Confidence among Rural
      Entrepreneurs in Finland
      Petri Ruuskanen   241

      Index   257

Figures and Tables

Figures

7.1 Twelve chicken barns near Normangee, Texas   107
8.1 Share of sales on retail distribution   124
10.1 The utilization of the wine crop by selected co-operative
cellars, 1996   161
11.1 Hitra-Frøya region and Røros region, Norway   177
15.1 Categories of Trust and Confidence   252

Tables

4.1 Major fair trade commodities and production
characteristics   64
7.1 Responses to the question: what exactly has changed to
make your community more or less desirable than in
1997?   112
10.1 Good wine production (million litres) by selected co-
operatives, 1990–1996   160
10.2 Relative profitability of wineries as measured by earnings
before interest and tax (per cent), 1996   161

Notes on Contributors

Reidar **Almås** is Professor of Rural Sociology, and Director of the Centre for Rural Research, Norwegian University of Science and Technology in Trondheim, Norway. He has published many books in the areas of rural development, farm policy and rural sociology. He is coordinating a national project to write the agricultural history of Norway and is currently writing the last volume (1920-present).

Daniel L. **Argo** is a graduate student in the Sociology Department at Sam Houston State University in Huntsville, Texas, US. His research area is the globalization of economy and society with a special emphasis on the informalization of work. He has presented several papers at professional meetings on this, and related, topics.

Dr Maria Rita Pontes **Assumpção** is a Senior Lecturer in Industrial Engineering, University Federal of São Carlos, Brazil. She has published articles in *Logistics and Restructuring of Agriculture* and *Supply Chain Management.*

Alessandro **Bonanno** is Professor and Head of the Sociology Department at Sam Houston State University in Huntsville, Texas. His research focuses on the implications that globalization has for democracy, including emancipatory options for subordinate groups. He has published 11 books and more than 60 refereed journal articles which have appeared in English and other major languages. He is currently First Vice President of the International Rural Sociological Association.

Hugh **Campbell** is a Senior Lecturer in Social Anthropology in the School of Social Science, Otago University, New Zealand. He is also the Director of the Centre for the Study of Agriculture, Food and Environment at Otago University. His research interests include: the development of organic agriculture in New Zealand, global trade and sustainable agriculture, the impact of neoliberalism on agriculture in New Zealand, the sociology of food scares, and rural masculinities.

Caron **Cates** is a graduate student in the Sociology Department at Sam Houston State University in Huntsville, Texas, US. Her research area is community quality of life issues related to industrial forms of agriculture. She has presented several papers at professional meetings on this topic.

Douglas H. **Constance** is Assistant Professor in the Sociology Department at Sam Houston State University in Huntsville, Texas, US. His research focuses on the socio-economic impacts of the globalization of the agro-food sector. He has co-authored a book on this topic and his work has appeared in journals such as *Critical Sociology, Rural Sociology, Agriculture and Human Values, International Journal of Sociology of Agriculture and Food, Culture and Agriculture,* and *Research in Social Movements.*

Joachim **Ewert** is lecturer at the University of Stellenbosch, South Africa. He has published in the international journals *Sociologia Ruralis* and the *Journal of Peasant Studies.* He is currently working on a book on the South African wine industry, jointly with Gavin Williams of St. Peter's College, Oxford.

Ivaldo **Gehlen** is Professor in the Sociology Department and Vice-Coordinator of the Post-Degree Program of Sociology at the Federal University of Rio Grande do Sul, Brazil. His research focuses upon social forms of agriculture, social-professional identity of farmers and the social impacts of technological changes on the agro-industries food chain - with a special emphasis on competition and sustainability. He has published on this, and related topics, and also on farmer social movements.

Associate Professor Erica **Hallebone** lectures in research methods in the School of Management, RMIT University, Melbourne, Australia. Her recent research publications are in the fields of globalization, technology and social impacts. She has conducted research for government, and has published articles on the social impacts of gambling, socio-demographic decline in rural Australia and the forms and experiences of personal identity.

Mirenda **Harris** is a graduate student in the Sociology Department at Sam Houston State University in Huntsville, Texas, US. Her research area is community quality-of-life issues related to industrial forms of agriculture with an emphasis on gender issues. She has presented several papers at professional meetings on this, and related, topics.

Geoffrey **Lawrence** is Professor of Sociology and Head of the School of Social Science at The University of Queensland, Brisbane, Australia. He is co-editor of the *International Journal of Sociology of Agriculture and Food*. His latest co-authored/co-edited books are: *A Future for Regional Australia: Escaping Global Misfortune* (Cambridge, 2001); *Altered Genes: the Future?* (Scribe, 2001) and *Environment, Society and Natural Resource Management* (Edward Elgar, 2001).

Mary **Mahoney** is a Senior Lecturer in the School of Health Sciences at Deakin University, Melbourne, Australia. Her research interests focus on impact assessment including the cumulative impacts of policy changes and the impacts of sport/physical activity in fostering social capital, community cohesion and regeneration in small rural communities. She is currently exploring the application of health impact assessment as a framework for developing healthy public policy.

Philip **McMichael** is Professor and Chair of Rural and Development Sociology, Cornell University, Ithaca, US. His research interests include global political economy and food security issues. He is the author of *Development and Social Change: A Global Perspective* (2000) and is past President of the Research Committee on Agriculture and Food of the International Sociological Association.

Mara **Miele** is a Lecturer in Marketing of the Agrifood Sector, and Agricultural Economics and Policy, at the University of Pisa, Italy. Her research interests include culture of consumption, consumer behaviour towards food, organic farming, animal welfare issues, and rural development. She is the author of *Atteggiamento dei Consumatori e Politiche di Qualità della Carne in Italia e in Europa* (with Vittoria Parisi), and *La Commercializzazione dei Prodotti Biologici in Europa*.

Manuel Belo **Moreira** is Professor at the Department of Agricultural Economics and Rural Sociology at the Higher Institute of Agronomy, Technical University of Lisbon. He has published in many international journals including *Agriculture and Human Values* and the *International Journal of Sociology of Agriculture and Food* (of which he is a former co-editor) and has written numerous chapters in books on the political economy of agriculture. He is President of the Research Committee on Agriculture and Food of the International Sociological Association.

Jonathan **Murdoch** is a Reader in Environmental Planning in the Department of City and Regional Planning at the University of Cardiff, UK. His research interests include social and spatial theory, planning and governance, food sector analysis, and rural development. He is the co-author of both *Constructing the Countryside* and *Reconstituting Rurality*, and has published in the journals *Sociology, Sociological Review, Sociologia Ruralis, Society and Space, Economic Geography*, and *Progress in Human Geography*.

Laura T. **Raynolds** is an Associate Professor of Sociology at Colorado State University, US. Her research focuses on globalization, agrarian restructuring, fair trade, and gendered labour forces in the Caribbean and Latin America. Recent articles have appeared in *Agriculture and Human Values, Gender and Society, International Journal of Sociology of Agriculture and Food, Human Organization, Global Development Studies* and *World Development*.

Javier **Rodríguez** is Assistant Professor of Economics at the University of Buenos Aires. He is currently working in the area of the globalization and transformation of Argentina's agro-food systems and their impacts on society.

Petri **Ruuskanen** is a Research Fellow in the Department of Social Sciences and Philosophy, University of Jyväskylä, Finland. He has published several articles and monographs on entrepreneurship, trust and the transformation of Finnish rural policy in the globalizing age.

Mark **Shucksmith** is Professor of Land Economy and Co-Director of the Arkleton Centre for Rural Development Research at the University of Aberdeen, Scotland. He is also programme adviser on rural issues to the Joseph Rowntree Foundation, expert adviser to the Scottish Parliament's Rural Affairs Committee, and Program Chair for the 2004 World Rural Sociology Congress in Trondheim, Norway. Research interests include social exclusion in rural areas, rural development, agricultural change, and rural housing.

Miguel **Teubal** is Professor of Economics in the Economics and Sociology Department of the University of Buenos Aires, and researcher for the National Council for Scientific and Technological Research (Argentina). He has worked on the political economy of Argentina and Latin America with special reference to food and agro-food systems and has published several books and articles on these topics.

Mardie **Townsend** is a Senior Lecturer in the School of Health Sciences at Deakin University, Melbourne, Australia. Her research interests focus on social and environmental sustainability, participatory democracy, and rural community health and wellbeing, including the cumulative impacts of policy change on rural communities, the measurement of social capital in rural communities, and the role of sport and physical activity in maintaining social capital in rural communities.

Acknowledgements

The genesis of this book was the Tenth World Congress of Rural Sociology held in Rio de Janeiro, Brazil, in July 2000. The theme of the Congress was *Sustainable Rural Livelihoods: Building Communities, Protecting Resources, Fostering Human Development.* Our major role at the conference was to organize the workshop on 'local responses to global integration' in which over 30 papers were delivered. It was clear to us that the material presented – by sociologists from 18 different nations – provided an excellent comparative dimension to the forces, and local impacts, of globalization. Following the Congress we invited a representative group of presenters to contribute chapters to a book on globalization/localization. Chapters were written, edited, re-worked and submitted for publication over the subsequent 18-month period.

While the book emerged within a framework of strong global collaboration among sociologists, we nevertheless faced difficulties. The main one was that of language. We wanted to provide an opportunity for colleagues in non-English speaking nations to bring their research findings to an English-speaking audience. We did not realize at the time the difficulties we would face in translation, and the problems of electronic submission of materials for publication. We thank all contributors to the book, but particularly the Spanish and Portuguese-speaking contributors for their patience and for their determined efforts in completing the chapters. Our Spanish and Portuguese-speaking colleagues were aided by Janet Norton who not only provided assistance with translation, but also undertook the organizational tasks of liaising with authors, formatting early versions of chapters, reviewing chapters for style and referencing, and keeping all contributors abreast of the book's progress.

Dimity Lawrence typeset the camera-ready copy of the book - correcting for style and readability along the way - and assisted in the compilation of the index. We thank the organizers of the Tenth World Congress of Rural Sociology for having provided an important opportunity for some excellent North/South dialogue in relation to issues of global/local interactions. We also thank Mark Shucksmith who helped to coordinate the workshop, and Manuel Moreira who provided strong support to the Portuguese-speaking delegates. A reworked and updated version of Philip McMichael's (2000) paper 'The Power of Food' (*Agriculture and Human Values*, vol. 17, pp. 21-33) has been prepared for this collection, with the kind permission of Kluwer Academic Publishers. We are grateful to Central Queensland University's Institute for Sustainable Regional Development for providing the funding for editorial assistance. Finally, we thank the staff at Ashgate for their patience and advice during the preparation of the manuscript. Particular thanks goes to Liz Barnsdale, Assistant Desk Editor with Ashgate, who reviewed the draft of the book and alerted us to necessary changes. We hope that readers will find the book's contents to be as insightful and important as the publisher and editors believes them to be!

We dedicate this book to those peoples – around the globe – involved in the struggle to create sustainable livelihoods for themselves, their families and communities.

Reidar Almås
Geoffrey Lawrence

PART I
GLOBALIZATION AND FOOD

Chapter 1

Introduction: The Global/Local Problematic

Reidar Almås and Geoffrey Lawrence

The Meanings of Globalization

The nature and extent of the impacts of global processes on local communities and spaces is occupying the minds of researchers, from a variety of disciplines, throughout the world. Is globalization destroying local social systems? Is it creating opportunities for economic growth? Does it impact positively or negatively upon the natural environment? Is the nation state in a position to mould global forces to the advantage of its citizens, or is it complicit in the expansion of neoliberalism to enhance profit making among the corporate elite? Finally, can globalization be harnessed for the good of local people, and if so, how? These are but some of the issues being discussed by academics, policy makers and, perhaps more importantly, by members of local communities, in an effort both to understand, and to seek to derive benefits from, globalization.

For the purposes of this book we define globalization as a process through which time and space are 'compressed', as new technologies, information flows, trade, and power relations allow distant actions to have increased significance at the local level (Gray and Lawrence, 2001). We are observing, in other words, an increasing intensity – a widening, deepening and speeding up – of global and regional interactions (Held *et al.*, 1999). What appears to be missing is a critical appreciation of the consequences. What we address in this book is the degree to which global processes are impacting upon local communities, the ways those impacts are occurring, and the extent to which global processes provide – or fail to provide – a basis for sustainable livelihoods, particularly in rural and regional communities.

In the early 1990s Ross and Trachte (1990, p.223) identified the activities of what were prototypical global firms – ones which viewed regions of the world as a 'mosaic of differentiated sites of potential investment' and which, because of their mobility, were actively disciplining both labor and the state. The authors recognized a reluctance on the part of the state to impose new taxes on these transnational firms, just as they recognized the state's retreat from so-called welfare state policies in the face of possible declining investment from those companies.

Others, such as Drache and Gertler (1991), identified the creation of a policy environment in the 1990s in which businesses were able to move their capital around the world with minimal restriction, and where deregulation of the labour market was being viewed as crucial by both firms and by nation states as a basis for continued capital accumulation. Firms were competing globally and, as such, adopted new sophisticated and productivity-generating technologies as well as seeking ways to minimize labour costs and improve profits.

Teeple (1995) argued that a global regime had emerged from an earlier 'international regime'. The international regime of capitalism had its beginnings in 16th Century trade between nations. Up to the Second World War such trade was reliant upon national capitals, controlled by nation states. The period from the Second World War to the 1970s represented an 'interregnum between the age of competing imperial powers and the coming of the global economy' (Teeple, 1995, p.57). The international regime following the war was one based on US hegemony – linked to the expansionary needs of US corporations. With the development of economies outside the US, there was a growing interdependence between nations, and between the branches of companies located within, and outside, specific nations. Regulatory and other mechanisms were set in place to liberalize world trade, contain socialism, promote legislation favourable to capitalist expansion, and to aggregate world markets.

The formation of the EC, NAFTA and other recent trade groupings can be viewed as part of market aggregation, something considered by Teeple (1995) as having been crucial for the promotion of free trade regimes in the 1980s and beyond. But global trade also represented a threat to the sovereignty of the state. The growth of corporate power had provided enough strength for capital to demand the dismantling of national barriers to trade. The growth of foreign capital and currency markets meant that economic decisions were being undertaken outside the context of the nation state. A 'global' system was seen to have evolved after 1970. Its features were the removal of tariff – and other – barriers to world trade, and the demise of US 'paramountcy' as corporate entities from Japan, Korea and other nations began their global spread in search of profits (Teeple, 1995).

It is undeniable that the growth in transnational corporations (TNCs) has continued apace since the 1970s. During the 1970s about 7,000 TNCs were in operation, but by the mid 1990s there were over 39,000 parent companies (with control over some 270,000 affiliates) and by the end of the century there were some 60,000 parent companies controlling 500,000 foreign affiliates. It is estimated that the parent companies and their affiliates contribute 25 per cent of the value of total world output (Paddon, 2001, p.98) and about two thirds of world trade (McMichael, 2000a, p.95), with the market value of individual corporations being greater than many nation states.

The corporations have understood well the profit making opportunities of intrafirm global trade. As McMichael (1995, p.90) has noted:

> TNCs subdivide production sequences according to technological or labor skill levels. Moving labor-intensive activities to Third World export platforms is routine. Typically, high technologies remain monopolized by First World firms, with component processes (assembling, etching, and testing computer chips), component goods (pharmaceutical stock, engines, auto parts), and consumer goods (cameras, electronic games, TVs, and video-recorders) moved offshore for production in the Third World.

For McMichael (2000a) the so-called 'globalization project' is a coherent approach to corporate profit making, creating conditions where trade is organized on a world basis by a largely unaccountable political and economic elite. The latter groups act as a power bloc to ensure that producers and consumers are connected, across space and time, within a global marketplace – one increasingly regulated by supranational entities, such as the World Trade Organization (WTO). A neoliberalist discourse usually accompanies the new pattern of investment. It is one which highlights the social benefits of self-regulating markets, endorses the theory of comparative advantage, and lauds the prospects for entrepreneurialism to bring benefits to local economies (Gray and Lawrence, 2001). Its promise is the delivery of the products of global industry to all peoples, and the extension of democratic principles across the entire globe. But is this happening?

Capitalist development has been, and continues to be, uneven development (Held *et al.,* 1999) and while some areas might be prospering, others are not. Communities are assigned particular roles in the global economy, with the orchestration of product manufacture and distribution occurring centrally by corporations and with local regions having to perform particular tasks as part of the overall global plan. Flows of finances and products have allowed TNCs to flourish at the same time as they have increased the economic instability – and vulnerability – at particular sites of production. The threat of capital withdrawal, in the context of the lack of control by the nation state over foreign firms, has eroded national sovereignty over domestic policy (Hamilton, 1996). For some regions (such as Asia in the 1970s and 1980s) the combination of profit-seeking TNCs in concert with a compliant state has meant the 'unlocking' of relatively cheap labour, and with it a raw and unforgiving form of economic development, often accompanied by environmental degradation (see Beck, 2000; Hamilton, 2001). For others it has meant the withdrawal of capital from economic activity once sanctioned under a corporatist mantle that saw labour engage with capital on a much stronger footing (McMichael, 2000a). Here, accompanying the removal of legislation that once provided for tripartite (business, labour, state) decision making in economic matters, has been the flight of capital – leaving workers with several choices: accept reduced wages and/or increase productivity, or lose jobs. Protest has also been an option, albeit one that has not necessarily advantaged the workers (see McMichael, 2000a). As Jones (2001, p.13) and Gray and Lawrence (2001) have suggested, the typical policy mix adopted by states seeking to capture the transnational investment dollar has included:

- Internationalization of the economy via the lowering of barriers to foreign trade and investment;
- Deregulation and privatization of sectors formally controlled by the state;
- A movement away from Keynesian macro-economic policy to one of controlling inflation via monetary policy;
- A greater trust of unfettered market forces to resolve issues of resource distribution;
- Changes to industrial relations systems enhancing individual contracts and marginalizing unions;
- A winding back of 'welfare state' policies that provided both an improved 'social wage' and a safety net in the event of sickness and unemployment; and
- The lowering of taxes on corporate profits, capital gains and high incomes along with the implementation of 'user pays' principles for state-based services.

Jane Kelsey's (1995) account of the so-called New Zealand Experiment implemented by a Labour government in 1984, and continued by a conservative National government throughout the 1990s provides dramatic evidence of the social and economic malaise that accompanied the policies above. Stagnation, recession, growing unemployment and social polarization were the main outcomes of policies aimed, ostensibly, at forging transnational connections with the New Zealand economy. Epp and Whitson (2001) argue that similar policies in Canada have been responsible for what they term the 'writing off of the rural west', while Pritchard and McManus (2000) describe contemporary rural Australia as a 'land of discontent' as a consequence of the destructive forces of rural restructuring under neoliberalism. As Jones (2001, p.13) has commented:

> In many countries, such measures served to shift socio-economic patterns of income, wealth, and life chances from a diamond shape (with a bulge in the middle signifying a large middle class) towards an hourglass shape (representing growing segments of haves and have-nots and a shrinking middle class).

The New Zealand economy, like that of many other nations adopting the economic rationalist policy framework described above, certainly received the foreign capital investment that had been anticipated. Despite its misshapen form, investment capital flowed in, increasing foreign ownership (Kelsey, 1995). But what of those nations unable to persuade capital to move to their shores? Castells (1996) has employed the telling phrase 'structural irrelevance' to describe those regions (such as many of the rural regions of nations such as India, China and Latin America) which have no likelihood of securing global capital. For Castells (1998) such structural irrelevance will be worse than any economic dependency that might have resulted from global investment. Hoogvelt (1997, p.240) describes the fate of those unable to enter into global circuits of investment and production:

> Performing neither a productive function, nor presenting a potential consumer market in the present stage of high-tech information-driven capitalism, there is, for the moment, neither theory, world view, nor moral injunction, let alone a programme of action, to include them in universal progress. Developmentalism is dead, containment and exclusion rule OK!

It is the future of these regions which will, no doubt, occupy the minds of activists, planners and community members over coming decades, particularly if the outcome of containment and exclusion is political anarchy (Hoogvelt, 1997), the weakening of democracy (Beck, 2000) and the growth in the militancy of marginalized groups (Held *et al.,* 1999; Waterman, 2001).

Global Meets Local

Perhaps as a result of the nebulous nature of the catchall term 'globalization', perhaps because of the complex nature of the social and economic arrangements that foster global/local interrelations, or perhaps because of the inability of social scientists to grasp the multiplicity of social forces that are both driving, and responding to, globalization, it remains that there is no clear understanding of the relationship between global change and local community viability. It is difficult to generalize. What can be stated, however, is that patterns of global/local interaction are changing and that we can trace such changes through an analysis of concrete social relations as evidenced in such things as employment and working conditions, industrial restructuring, local culture, and the environment.

Working conditions and industrial restructuring have been discussed briefly above. There is evidence that as globalization proceeds, wage workers experience job intensification and employment insecurity (Hoogvelt, 1997). According to McMichael (2000a) as transnational capital moves to low wage zones of the global economy, downward pressure is placed on workers' salaries and benefits in other, more developed (and unionized), regions. The sequence is as follows: policies are put in place to entice capital to invest – usually meaning state concessions to capital including mechanisms to lower wages; lower, or stagnating, wage levels decrease purchasing power and reduce the capacity of the state to raise revenue; there is a reduction in public expenditure thereby removing part of the platform of support that workers and communities have won in previous struggles; wage workers experience a standard of living decline at the same time as corporate capital's export expansions bring profits to shareholders. If and when the company finds it is more profitable to move to a cheaper location, it does so. In the meantime, it takes the opportunity of reminding the workers that they should not complain – at least they are in paid employment! This is part of the so-called 'race to the bottom' scenario for workers.

We should recall, however, that when jobs are exported to developing nations, there can be considerable benefits to those recipient regions. Such benefits include the challenge to feudal economic and social relations that impede the spreading of benefits to the poor (Seitz, 1995), increased employment opportunities, investment in local infrastructure and skill-based training. Women can enter the workforce, improving family income and women's status (Mittelman, 1996). The WTO-promoted dismantling of trade barriers has the potential to give developing nations greater access to markets in the developed world (Legrain, 2001). Foreign capital can lean upon recalcitrant governments to improve education and to provide housing for workers. Expenditure by those in jobs can generate local industry – particularly knowledge intensive services (see Kanter, 1995). It is when paid work takes the form of exploitation and there is a clear violation of human rights that questions are raised about the supposed benefits that corporate investment brings (Barnet and Cavanagh, 1994). The question of the real beneficiaries of globalization must also be asked. As Legrain (2001, p.43) asserts:

Free trade is indeed a wonderful thing; it is a pity that rich countries do not practise what they preach. Not only do rich countries conspire to keep out poor countries' main exports, agriculture and textiles, they are also busy carving up world markets through preferential pacts that make a mockery of free trade.

In relation to local cultures, one of the main issues raised is the extent to which they are being Americanized and/or homogenized. According to Featherstone *et al.*, (1995, p.2) the global era is one in which 'flows' (of people, ideas, finance and technologies) have greater centrality to economic and social life than do national institutions. For some authors there is a 'sameness' to the styles adopted by the global media, similarities in the convenience foods prepared and eaten all over the world, and widespread purchase of the products of mass production (Barnet and Cavanagh, 1995). There are direct similarities between the sorts of cars driven, television and radio sets switched on, satellite and computer technologies accessed, management regimes promoted, and recreational and leisure pursuits adopted. But while state borders might be being rendered obsolete, does the presence of American and other 'western' goods, services and ideas mean that there is an inevitable standardization occurring throughout the world?

For some writers, such as Ritzer (1996) the McDonaldization of the globe is representative of the emergence of homogenous 'one world' culture. Standardization – viewed as a necessary consequence of the production-process mentality that infuses capitalist organizations and seen, in particular, in the fast food industry – occurs as instrumental rationality becomes the organizing principle of modern life. Following Weber, Ritzer argues that when firms seek to maximize efficiency, and to enforce calculability, predictability and control, the outcome is an inevitable bureaucratization that, spread around the world, leads to routinization and standardization. (The McDonaldization thesis is further discussed by Miele and Murdoch in Chapter 2 of this book.)

Perhaps more subtly, McGrew (1992) believes globalization both 'stretches' and 'deepens' social interactions. That is, its political, socio-cultural and economic effects are being felt around the world in a more profound (deeper) way with the result that people, increasingly, come to see themselves as global citizens. We should remember that much of the content of popular culture, which is consumed throughout the world, is produced by people in those nations most closely associated with globalization – with advertising and movies clearly depicting the lifestyles and attitudes of the west. Cultural flows move around and under state border controls. While certain that people have individual choice, Waters (1995, p.149) nevertheless argues:

the mass media knit the global culture together by means of content as well as by means of language. They do this not merely by offering common simulation opportunities but by magnifying global problems and global events...we now look at the world through global spectacles...(Media events) are deliberately constructed as stylized mass entertainments and they are, in Durkheimian terms, collective representations of global commitments to democracy, consumption, capitalism and a liberal tolerance of diversity.

There are many authors who disagree with Ritzer and the convergence theorists. Robertson (1995), for example, was so concerned that globalization would be viewed, uncritically, as an homogenizing force which would lead to convergent development that he employed the term 'glocalization' to remind us that global forces do not override locality and that homogenization and heterogenization are both crucial features of modern life. He was also concerned that globalization would become a 'grand narrative' in an era where the postmodern condition should warn us against employing such narratives. In contrast to Ritzer, Robertson argues that we do not live in a 'McWorld of homogenizing globalization', but one in which, to use a phrase of Jean-Paul Satre's, local and regional areas are 'condemned to freedom' (see Robertson, 1995, p.39). Localities are forced to interact with wider forces to maintain their economic future. But they are able to do so in very different ways and under a variety of conditions. In this formulation, the local is not the 'other' of the global. With time compression of the global economy, the local – in a dialectical manner – absorbs, shapes, alters and opposes wider tendencies while creating and promulgating its own. These, in turn, contribute to the eventual shape of macro or global forms.

Like Robertson, Pieterse (1995) rejects any view that globalizing processes are necessarily making the world more uniform and standardized. There may be some synchronization brought about by technological, economic and cultural flows, but the effect is hybridization where cultural forms become separated from current practices and are recombined in new practices (Pieterse, 1995, p.49) forming not a global whole but a global 'mélange' of hybrid sites and spaces. Pieterse believes it is possible to replace an older 'homogenizing' vocabulary with a new one. Instead of modernization we have (many) modernizations; instead of Westernization as a unidirectional force, we have global mélange; instead of cultural synchronization we are seeing 'creolization' and crossover; and instead of a totalizing world civilization we can talk, instead, of a global ecumene (Pieterse, 1995, p.62).

How, then, does one understand the McDonaldization of society? Surely the existence of this form of enterprise in most countries of the world suggests increased uniformity and standardization? The answer would seem to be yes and no. There is a paradox. Consumers are able to exercise choice (by not going into a fast food restaurant). They have free will and are certainly not 'cultural dupes'. But, to revisit Ritzer, while the consumer is able to exercise choice, the reorganization of consumption represented by the ever-growing sale of mass foods, and the 'rationalization' of food preparation which once took place in the domestic sphere, are viewed as diminishing the choices available. In other words, consumers are free to choose but many, in the context of the assault upon local culture by global firms, appear happy to choose uniformity and standardization! It is here that – although largely beyond the scope of this book – an understanding of the hegemonic power of the corporate media in concert with the household logic of cheaper and more convenient food purchases needs to be explored.

Global also meets local at the level of the environment. What are the environmental impacts of corporate profit making? The sophisticated industrial processes that lead to the mass production of goods and services are creating waste and pollution on an unprecedented scale. Greenhouse gas emissions, ozone-depleting chemicals and toxic agri-chemicals remain as perhaps the most insidious forms of planetary destruction. It is estimated that one-fifth of the world's population breathes poisonous air – that is, above the acceptable levels identified by the World Health Organization (Goodland, 1996, p.208). There are numerous examples of today's production regimes severely compromising the environment. Agri-business firms continue to manufacture and distribute pesticides, herbicides, fungicides and a variety of 'chemical fixes' that have a major impact on soils and waterways (Altieri, 1998; Foster and Magdoff, 1998). Indeed, about 35 per cent of the earth's land is already degraded from soil erosion, salination and desertification – outcomes of what has been termed 'productivist' agriculture (see Vanclay and Lawrence, 1995; Goodland, 1996; Gray and Lawrence, 2001). The packaging and movement of foods over vast distances through agri-food chains is energy intensive and while providing convenient and cheap foods for some peoples (those in the developed world) is nevertheless premised on the creation of agro-food export regions which may disadvantage other peoples (in the developing world) (Foster and Magdoff, 1998; McMichael, 2000b).

The polluting activities of mining companies in developing nations such as Indonesia and Chile continue unabated (Hamilton, 2001), while the deforestation occurring in nations of Central American, and in countries such as Borneo and Brazil is believed to be occurring at an unprecedented rate. According to Seitz (1995, p.155) over the past 100 years developing nations have lost over half their forest cover, leading to soil erosion, the 'hardening' of the soil profile, and to the flooding of regions which had previously never experienced flooding. People have been dispossessed of their land, homes and livelihoods. Acid rain is another feature of modern production and its effects are widespread. As greater amounts of fossil fuel are burnt, the resultant gases mix with moisture, oxygen and sunlight to produce sulfuric and nitric acid. This so-called acid rain falls in a non-specific pattern, corroding the built environment, killing fish and trees and retarding the growth of crops. Its affects on humans has yet to be quantified, although it is known that chemical-infused smog has significant health affects for humans including asthma and other respiratory problems (Seitz, 1995, p.144). Finally, biodiversity is being threatened as many of the above processes singly and together impact upon plants, animals and other life-forms throughout the world (Harding, 1998).

There are other specific and well-known incidents that are associated with the negligence – or at least with a non-caring attitude of – the corporate sector. The deliberate dumping by the Hooker Chemical Company of over 20,000 tons of wastes into New York State's Love Canal during the 1940s and 50s led to the contamination of over 600 homes and to a myriad of illnesses decades later (Seitz, 1995, p.153). Nothing much has changed: during the mid-1990s the US alone produced one trillion pounds of toxic waste every year, much of which is still not disposed of in an adequate manner (Seitz, 1995; Hawken *et al.*, 1999.) The grounding of, and subsequent oil spill from, the Exxon Valdez in Alaska in 1989 is unlikely to have occurred if Exxon had not reduced the staff levels on supertankers by one third and 'forced' its ships through faster, but more dangerous, shipping lanes (Hawken, 1993; Goldsmith, 1996). Union Carbide's Bophal accident in India in 1984 when the release of methyl isocyanate gas poisoned over 200,000 people (of whom 6,000 died) is seen to have resulted from the company's cost-cutting measures, which increased risk at the local level (Goldsmith, 1996, p.88). On the island of Bougainville, BHP-Billiton's Ok Tedi copper mining venture poisoned and flooded rivers leading to rebellion which resulted in the deaths of thousands of Indigenous people (Harding, 1998). More recently the leaking of chemicals into world heritage listed areas (as is occurring in Australia with agri-chemicals into the Great Barrier Reef, and uranium into the Kakadu National Park) is undermining the hard-won battle waged by environmentalists to prevent industrial pollution in ecologically pristine areas. 'Mad cow' disease (bovine spongiform encephalopathy) is unlikely to have occurred in the absence of the profit making zeal of an industrialized agriculture that condoned the suspect practice of feeding animal offal to herbivores. We also need to be aware of the environmental impacts of tourism (Barnet and Cavanagh, 1995), of dam building (Walker and Crowley, 1999), biopiracy and the appropriation of genetic material from developing nations (Christie, 2001) and of the loss of biodiversity as a direct result of unsustainable farming practices (Dovers, 1999; Gray and Lawrence, 2001). Many of these problems are a consequence not only of the activities of capitalist firms, but also of the pro-development policies enacted by compliant states.

Sustainable Livelihoods

Take one of the problems mentioned above – that of deforestation. According to Vandana Shiva (see Beder, 1996) the loss of timber from the Indian countryside (and its increasing unavailability to local people) is not caused by 'exploitation' of the resource by local people, or by 'overpopulation'. Rather, it can be explained by the displacement of people from the commons, and the harvesting of timbers from what were natural forests for commercial, export, dollars. The tendency to assign blame to the victim is a widespread one. Local people cannot hope to progress unless the conditions for social and economic reproduction are in place and the activities of corporate firms are in accord with the so-called 'triple bottom line' where a balance is struck between economic growth, social progress and environmental security. For this to occur ways must be identified to empower citizens to act to create new resource-use trajectories – those which will contribute to sustainable livelihoods.

The idea of sustainability is a powerful one as it has the potential to alter the ways that production occurs. It captures the idea that we must work toward satisfying the needs of the current generation, without endangering the lives of future generations (Sachs *et al.*, 1998). It takes as its premise that society will advance when economic growth and environmental protection are complementary (Dryzek, 1997; Hawken *et al.*, 1999) rather than being at odds with each other. The World Conservation Strategy (see Venning and Higgins, 2001, p.31) defines the principles of sustainable development as those of:

- The integration of conservation and development;
- The satisfaction of basic human needs;
- The achievement of equity and social justice;
- The provision of social self-determination and cultural diversity; and
- The maintenance of ecological integrity.

There has been some difficulty in operationalizing these principles. What must be understood, however, is that sustainable development is more encompassing than 'ecologically sustainable development'. As the dot points suggest, sustainable development posits an intimate interconnectedness of social, economic and environmental processes. While some writers have seen sustainable development as an oxymoron and point to the glee with which big business embraces the term as part of its 'greenwash' (Beder, 1997), others have endorsed the adoption of the principles and practices of sustainable development as perhaps the only key to prosperity for humanity (Hawken *et al.*, 1999).

Questions remain, of course, about how any increased consumption of goods and services could be sustained – given the extent of today's pollution and resource use (Beder, 1996). There are concerns about the extent to which sustainable development is an ideological and political tool to keep development out of the Third World, and yet allow First World consumption to continue unabated (see Beder, 1996). There are also – for obvious reasons – misgivings about trusting companies that have been major polluters with the task of cleaning up the environment (Kloppenburg, 1991; Beder, 1997). There are also concerns about the specific tools that might be employed by business to help in the task – biotechnologies deserving special mention (Ho, 1998; Hindmarsh and Lawrence, 2001). Finally, there are the questions of how equity, social justice, self-determination and cultural diversity can be 'sustained' in the face of the erosion of the citizenship, and the processes of detraditionalization, in many countries (Held *et al.*, 1999; Gray and Lawrence, 2001).

In this book we employ the phrase 'sustainable livelihoods' as we believe it both overcomes the possible problem of being too broad a term to be usefully applied (a criticism of 'sustainable development') and that it, more importantly, focuses attention upon the local, community-focused, dimension of sustainability. The term 'sustainable livelihoods' was used in the title of the Tenth World Congress of the International Rural Sociological Association's conference in Rio de Janeiro in 2000 and has been given specificity by de Haan (2000). The aim is to identify components of local communities that provide for long-term sustainability. These include social and individual capital, the natural capital base of the region, the institutional capital available to communities, and the financial capital that enters or leaves. While we have not 'imposed' such a framework on authors in this collection, there is, in virtually every chapter, a discussion of how local communities deal with, and either profit or lose from, their engagement with global forces and entities. This in turn, gives us important clues to the constellation of 'capitals' that are at play in fostering community development and ensuring that development at the local level is compatible with ecological integrity.

Social capital – the networks of trust and cooperation – is viewed as being an essential ingredient in sustainable regional development (Gray and Lawrence, 2001). Communities with a high degree of public interaction and civic pride are considered to have a better chance of survival than those where social capital is limited (see Almås, Chapter 11). Yet, global processes and neoliberalist government policies have been implicated in the depletion of social capital and with it growing social inequality and social exclusion (see Shucksmith, Chapter 13). How, then, might social capital be built at the local level? Some ideas – such as through the 'slow food' movement, the fair trade movement, community supported agriculture, bartering/alternative money, bioregionalism, agro-ecology, 'natural capitalism' and locally-based organic production systems – have become much more than fanciful prospects (see Mander and Goldsmith, 1996; Hawken, *et al.*, 1999; Brunckhorst, 2000; Lawrence *et al.*, 2001; Michelsen, 2001; Tovey and Blanc, 2001; Venning and Higgins, 2001). It is unlikely that we can create a future if that future has not been imagined, and our imaginations must take us beyond the boundaries of an arid, arguably undemocratic and globally-polluting capitalism to a world where localism is celebrated and sustainable livelihoods are practised in harmony with the natural environment.

Part I: Globalization and Food

The focus of Chapter 2, by *Mara Miele and Jonathan Murdoch*, is on the standardizing and differentiating cultures of food. The conventional story of change in the food sector conveys a sense that all the main trends in both production and consumption are converging, leading to standardization and homogenization. The authors show that there is emerging evidence that some trends and processes are moving in the opposite direction – promoting diversity and differentiation in food production and consumption. Firstly, they tell the story of increasing standardization using the illustration of 'fast food'. They show how the growth of this sector does indeed promote growing uniformity. However, they also show that fast food provokes a reaction: concerned actors seek to resist this form of globalization by highlighting that which is local and distinctive. Taking the case of Slow Food, an international consumer movement, they analyze how such a movement may promote tradition, typicality and local distinctiveness. Miele and Murdoch conclude by drawing attention to some of the harmful economic, social and environmental side-effects of fast food networks. They point to the Slow Food movement as an alternative, showing that regional taste and local diversity can be strengthened.

Chapter 3, by *Maria Rita Pontes Assumpção*, analyses the restructuring of the sugar supply channels for food production in Brazil. As global corporations from the food sector have entered the Brazilian market, changes have been made in the structure among companies downstream in the distribution channels. There have been growing product innovations; in their local operating units food-processing corporations have had to develop technological capabilities in order to adapt their new products to Brazilian tastes. Furthermore, biotechnology has made new modified products available, which has led to greater uniformity, faster production, and extended product lifetime. The industrial market thus demands that sugar accords with standards and specifications established by the new production processes. Through a process of aggregate value, sugar has gone from being a direct raw material to the food industry, with its purchase based solely on price, to becoming an intermediary product forwarded to its buyers through technical sales. Services are added to sugar to ensure its quality and conformity to the specifications of the food recipes in which it is to be used. Coordination of the sugar chain now requires a capacity for technical sales of liquid sugar (sucrose) and invert sugar (sugar-in-solution from a chemical reaction). This coordination is being led by the producers of industrial ingredients, strengthening the coordination in the hands of global companies downstream from the mills.

In Chapter 4, *Laura T. Raynolds* looks at the Fair Trade movement, arguing that it represents an important new challenge to the global agro-food system. This movement critiques conventional patterns of agricultural production, distribution and consumption. The Fair Trade networks create new market links between progressive consumers in the North and producers pursuing more socially and environmentally sound production in the South. Though the international movement of Fair Trade labelled products is quite young, this trade has grown dramatically in recent years. The world market for Fair Trade products is currently valued at US$ 400 million, with sales growing at 10 to 25 per cent per year. Roughly 60 per cent of the Fair Trade market comprises food items. This analysis suggests that Fair Trade initiatives have begun to create important new networks which span the widening local/global divide. By building networks of solidarity between agro-food producers and consumers, fair trade initiatives encourage the participation and empowerment of disadvantaged Southern producers. The evidence reviewed by Raynolds suggests that Fair Trade can indeed create alternative market networks, bypassing the corporate-dominated world agro-food system. The greatest challenge for Fair Trade initiatives may be in ensuring that they go beyond creating a parallel market in the North, to integrating more sustainable environmental and social practices in the South.

Philip McMichael shows in Chapter 5 how the current restructuring of world agriculture intensifies a global division of agricultural labour. Historically, trade in low-value temperate cereals and oilseeds has been dominated by the North, and trade in high-value products has been distributed increasingly to corporate agro-exporters, or their contract farmers producing in the South. For Southern states this is often an unstable trade, signalling a widespread subordination of producing regions to global production and consumption relations, organized by transnational food companies. Under these conditions, agriculture becomes less and less an anchor of societies, states and cultures, and more and more a component of corporate global sourcing strategies. It increasingly anchors a system of global profiteering in food products, a system in which food travels an average of two thousand miles from farm gate to dinner plate. This is an emerging process of centralization of agro-food capitals involving food chain clusters of firms that 'control the food system from gene to supermarket shelf'. However, McMichael points out that there are also counter-movements. Some of these are direct challenges to the corporate regime – such as consumer movements concerned with labelling, food safety and fair trade; farm worker movements concerned about the use of pesticides and about worker security; and farmer movements concerned about protecting agriculture from incorporation into the corporate-dominated world of agribusiness.

Part II: The Restructuring of Local Agriculture

Assessing some fifteen years of neo-liberal reforms in Australia and New Zealand agriculture, *Hugh Campbell and Geoffrey Lawrence* show, in Chapter 7, how these experiments were sold as a potential crisis solution – only to plunge agriculture into a full *structural* crisis. The results were ongoing instability and the inability of neoliberalism to 'fix' or stabilize capital accumulation in a new national or global regime. Under the previous regime, the state had intervened to underwrite the social and economic infrastructure of rural communities. In contrast, the move towards neoliberalism saw an abandonment of any 'rural policy'. The immediate effect of economic liberalization was a decline in farm revenues. The suddenness of deregulation made this particularly painful in New Zealand – but similar effects were also emerging in Australia with collapses in wool, wheat and beef prices. Falling farm revenues were accompanied by severe declines in land value. Continued environmental degradation is expected to continue as financially-stretched producers continue to 'mine' the soil, and overstock and overgraze, in the hope of improved prices in future years. Despite some suggestions that a 'successor regime' might be emerging, they contend that crisis and uncertainty are the main features of 'neoliberal farming' in the Antipodes.

In Chapter 7 *Douglas Constance* and his colleagues present a study on a Texan controversy about corporate chickens and community viability. Investigating Sanderson Farms, the seventh largest broiler producer and processor in the US, the authors show how this firm was able to receive establishment support from state and local political elites. However, because of reported negative environmental and social impacts of the chicken barns, the company was met with widespread community-based resistance. Concerns were stench, fly infestations, health concerns, lack of enforceable controls or constraints by any government agency, manure run-off and loss of amenity. As in other episodes of globalization, the Sanderson Farms' project operated well at the economic level but failed at the local level. The decoupling of economic expansion and community socio-economic well-being and the absence of legitimising mechanisms to address emerging problems, served as vectors to radicalize community response to corporate penetration. This is perhaps the most important conclusion to be reached from the analysis of this case study: the opening of local communities to globalization is occurring in the context of a lack of institutions capable of buffering the unwanted consequences of the growth of capitalism and for controlling its most powerful actors. This situation engenders fierce resistance and creates a contested terrain in which corporate power is resisted and redefined.

Miguel Teubal and Javier Rodríguez argue in Chapter 8 that the agro-food system of Argentina has been greatly transformed in recent decades. They analyze the overall impact of structural adjustment policies and globalization processes on the restructuring of the Argentine agro-food system and how this has affected agriculture and access to food for vast segments of the population. The Argentine system is subdivided into 14 agro-industrial complexes and changes carried out in recent years in each of these are considered. The impact this restructuring has had on the social actors involved in the agro-food system and their relations are also considered. A corollary of this analysis has to do with how these transformations have influenced access to food for the lower income levels of society, and the changing characteristics of the Argentine agricultural sector, namely the difficulties for the survival of small and medium sized farms. The authors argue that the various factors operating in the Argentine food system are creating an increasingly polarized and 'disarticulated' society by contributing substantially to unemployment and under-employment in all its forms, low wages and an increasingly regressive income distribution, amongst other factors.

In Chapter 9 *Ivaldo Gehlen* assesses the social impacts of the technological transformations of household milk production in the State of Rio Grande do Sul, Brazil. The producers who were part of this study are still perplexed at the transformations in progress. But they are politically organized and looking forward to sharing productive solutions: community coolers, rotating credit, bartering, technical assistance in groups, technology exchange, and micro-regional milk-belt planning. They systematically put forward public actions and struggles that demand mobilization and engagement. The result of these efforts is more dependent on the degree of the producers' mobilization and engagement than on exogenous technological factors. According to Gehlen, the modernization of Brazilian agriculture evidences its economic, social and ecological unsustainability. Agro-industries hold productivity as their priority while producers, organized in associations and cooperatives, prioritize social reproduction together with the agricultural workers' social, cultural, economic, and environmental sustainability. The adoption of alternative technological formats makes the social reproduction of the household agricultural workers possible, which enhances their inventive capacity. Organic production, increasingly accessible to these producers, tends to subvert the logic imposed upon the household producers by the big agro-industries and to constitute sites of resistance to this logic.

Joachim Ewert investigates, in Chapter 10, the integration of the South African wine industry into the global economy. Emerging from a highly-regulated past, South African growers, cellars and workers were thrown into a market-driven, globalized environment almost overnight. In this traumatic process, all stakeholders had to rely on their own resources, trying to survive as best they could, with little cooperation between them. Previously prohibited from selling their wine in overseas markets, a considerable number of farmers are profiting handsomely from the spectacular increase in exports, especially in the latter half of the 1990s. Although aggregate exports have levelled off over the last two years, growers who are able to deliver a 'quality' product have managed to remain profitable. However, the exposure to international competition also carries new risks. For the majority of farmers it has meant uprooting old vineyards and financing the cooperative in the effort to transform the cooperative both technically and organizationally. All this has to be done in a deregulated environment with no government subsidies. The people to benefit the least from South Africa's global integration are the vineyard workers. Whilst employment opportunities have increased since the mid-nineties, real incomes do not reflect the export boom. Paternalist consciousness, a lack of union organization and weak state capacity prevent workers from getting their share.

Part III: Communities and Resistance in a Globalizing World

In Chapter 11 *Reidar Almås* presents two different socio-economic strategies to meet the challenges from market and political forces in the age of globalization. The region of Røros, Norway, was formerly based on mining and mountain farming, while the resource base in the region of Hitra-Frøya was fishing and coastal farming. Today the coastal region is being integrated into the international fish markets through a highly competitive fishing and fish farming industry. The North Sea based oil industry is also making its impact, especially in the labour market, while agricultural production is rapidly decreasing. In the mountain region, mining is abandoned completely and the town of Røros has been converted into a tourist centre based on artefacts from mining culture, mountain farming and handicraft. Green tourism and eco-tourism is thriving on the proximity to small-scale mountain, and organic, farming. This small farming culture is struggling hard to survive under the pressure from deregulation and WTO demands. In order to survive, farmers are building coalitions with consumer groups in Norway. Why have the two regions adapted to globalization so differently? There are various explanations for these strategy choices largely relating to the regions' resource base, their former links to the world market, and to different cultural and political orientations.

Manuel Belo Moreira illustrates, in Chapter 12, some of the consequences resulting from the changes to nation-state regulations in the era of globalization. Global finance capital, through the world financial market, puts pressure on the nation-state. While we are used to seeing the state offering its 'facilities' to the highest bidder, under conditions of globalization the situation is reversed with transnationals auctioning – on an international basis – their availability and investment capacity, in a bid to determine which country can offer the best conditions to global capital. This changing role of the nation-state is one of the main features of globalization: each day it becomes less a regulator and more a facilitator of the requisites of capital. Withdrawing regulation functions from the nation-state have very uneven impacts in different socio-historic contexts. Especially for the rural areas, the abandonment of regulatory functions performed by the nation-states, or the lack of power to enforce them, is threatening the sustainable livelihoods of these rural societies. In terms of less favoured rural zones, the state is not only necessary in order to correct market failures; it is also essential to compensate for the lack of interest that market forces show in the future economic and social viability of these areas.

In Chapter 13 *Mark Shucksmith* discusses the ways in which economic, social and political forces for change operate in rural areas to produce social exclusion. He shows how the economies and societies of rural Europe are changing rapidly in the face of globalization, economic restructuring, and migration. These forces have different implications for different areas and different social groups. One key element is the increasingly global penetration of local markets. On the one hand, international capital may seek to exploit those rural areas characterized by low wages, a compliant, non-unionized workforce, and lower levels of regulation. On the other hand, local capital may seek to develop products that depend upon a local identity for their market niche, thus 'selling the local to the global'. Research into the disadvantaged in rural Britain identified labour markets and housing markets as instrumental in generating inequality and exclusion. Many respondents perceived very restricted opportunities for well-paid, secure employment or for affordable housing, while at the same time these markets enabled affluent households to move into rural areas. Young people and women tended to have the fewest options. These impediments to inclusion were closely connected to the failings of private and public services, most notably transport, social housing and childcare. To mitigate the failings of markets and the state, there was a greater reliance on the voluntary sector and on friends and family. Migration and the loss of young people have ruptured informal support networks and left elderly people socially isolated.

In Chapter 14 *Erica Hallebone, Mary Mahoney and Mardie Townsend* analyze the recent development of rural Australia. Over the past two decades, the population has declined in four out of every 10 rural shires in Australia, despite an increase in the national population in that period. In addition to the impacts of the globalization of agricultural markets and technical developments in farming, there is evidence to suggest that recent changes in public and private sector policies are compounding rural disadvantage. The authors studied two remote rural towns in order to see if attempts of rural development have a potential for success. Community development initiatives in these two rural areas are highly frustrated not only by external neglect but also by disintegrated policies and sources of assistance. Experiences within these communities show not only that strategies need to be put in place to assist them in enhancing the community sense of well-being and survival, but also that to be successful they need to be realistic about the future. To the extent that the trends of the agricultural industries – and the reduction of the welfare state and privatization – tend to lead to a loss of community identity and to an increase in individualism, the chances for building local initiatives are diminished. In the context of globalized capitalism and a national rural policy vacuum, the survival of communities such as these will depend increasingly upon on the initiatives of new local industries and the residents' construction of networks and social capital.

Finally, in Chapter 15, *Petri Ruuskanen* explores the problem of trust among rural entrepreneurs in Finland. The need for increased flexibility in most aspects of industrial production has been one of the central themes in economic debate in the industrialized countries. As a part of the debate, small firms and their networking have attracted the attention of governments' rural policies. It has been claimed that networking gives small rural enterprises a new competitive edge in the globalizing economy. Creating co-operative networks requires the building of trust and social capital. If trust is absent in social networks, all actors will sacrifice the potential gains of co-operation to the autonomous pursuit of self-interest. The article focuses on the problem of trust and social capital among rural entrepreneurs in Finland. Based on in-depth interviews conducted with self-employed entrepreneurs and small employers

Ruuskanen shows that rural entrepreneurs' attitudes toward networking and co-operation are ambivalent. Rural entrepreneurs are reluctant to participate in networks, because they often feel their structural position in the network is subordinate and insecure. Due to their lack of confidence, entrepreneurs avoid taking risks. They do not expand their businesses, but work long hours by themselves and rely on their family members. Consequently, they withdraw from joint activities. Instead of the motive of profit, their motives are often aesthetic, including such 'ideals' as freedom, independence, and autonomy.

Together the chapters show that globalization as such has no independent existence as a pure socio-economic process. Globalization must be understood as a complex sum of interdependent processes driven by finance capital.

For the purposes of this book, we have explored impacts at the local level. People experience that which is 'global' locally, in their neighborhood, at their workplace, in their organization. There are, indeed, losers and winners, as documented in the chapters that follow. But it is not necessarily predetermined who will be the losers and who will be the winners. And even though the transnational companies are well organized today, local communities possess an ability to act, to build networks and institutions locally, regionally and eventually at the global level. Social groups possess the ability to counteract the agencies and organizations that 'impose' a unidimensional transnational future. Where there is global economic action, there is local socio-political reaction. This book is only a starting point for our studies of these fascinating social battles and processes occurring in the 21st Century.

References

Altieri, M. (1998), 'Ecological Impacts of Industrial Agriculture and the Possibilities for Truly Sustainable Farming', *Monthly Review*, vol. 50(3), July-August, pp. 60-71.

Appadurai, A. (1990), 'Disjuncture and Difference in the Global Cultural Economy', in M. Featherstone (ed.), *Global Culture*, Sage, London, pp. 295-310.

Barnet, R. and Cavanagh, J. (1994), *Global Dreams: Imperial Corporations and the New World Order,* Touchstone, New York.

Beck, U. (2000), *What is Globalization?,* Polity, Cambridge.

Beder, S. (1996), *The Nature of Sustainable Development, Second Edition,* Scribe, Melbourne.

Beder, S. (1997), *Global Spin: The Corporate Assault on Environmentalism*, Scribe, Melbourne.

Brunckhorst, D. (2000), *Bioregional Planning: Resource Management Beyond the New Millennium*, Harwood, The Netherlands.

Castells, M. (1996), *The Rise of the Network Society*, Blackwell, Oxford.

Castells, M. (1998), *The End of the Millennium*, Blackwell, Oxford.

Christie, J. (2001), 'Enclosing the Biodiversity Commons: Bioprospecting or Biopiracy?' in R. Hindmarsh and G. Lawrence (eds), *Altered Genes II: The Future*, Scribe, Melbourne, pp. 173-186.

de Haan, L. (2000), 'Globalization, Localization and Sustainable Livelihood', *Sociologia Ruralis*, vol. 40(3), pp. 339-365.

Dovers, S. (1999), 'Institutionalising Ecologically Sustainable Development: Promises, Problems and Prospects', in K. Walker and K. Crowley (eds), *Australian Environmental Policy 2: Studies in Decline and Devolution*, UNSW Press, Sydney, pp. 204-223.

Drache, D. and Gertler, M. (eds)(1991), *The New Era of Global Competition: State Policy and Market Power,* Montreal: McGill-Queen's University Press.

Dryzek, J. (1997), *The Politics of the Earth: Environmental Discourses*, Oxford University Press, New York.

Epp, R. and Whitson, D. (eds)(2001), *Writing Off the Rural West: Globalization, Governments, and the Transformation of Rural Communities*, The University of Alberta Press, Edmonton.

Featherstone, M., Lash, S. and Roberston, R. (eds)(1995), *Global Modernities*, London, Sage.

Foster, J. and Magdoff, F. (1998), 'Liebig, Marx, and the Depletion of Soil Fertility: Relevance for Today's Agriculture', *Monthly Review*, 50(3), July-August, pp. 32-45.

Goldsmith, E. (1996), 'Global Trade and the Environment', in J. Mander and E. Goldsmith (eds), *The Case Against the Global Economy*, Sierra Books, San Francisco, pp. 78-91.

Goodland, R. (1996), 'Growth has Reached its Limit', in J. Mander and E. Goldsmith (eds), *The Case Against the Global Economy*, Sierra Books, San Francisco, pp. 207-217.

Gray, I. and Lawrence, G. (2001), *A Future for Regional Australia: Escaping Global Misfortune*, Cambridge University Press, Cambridge.

Hamilton, C. (1996), 'Economic Rationalism in the Bush', *Paper presented to the Department of Primary Industries and Energy*, Canberra, 19 November.

Hamilton, C. (2001), 'The Triumph of Ideology: Environment', in C. Sheil (ed.) *Globalisation: Australian Impacts*, UNSW Press, Sydney, pp. 187-201.

Harding, R. (ed.)(1998), *Environmental Decision-Making: The Roles of Scientists, Engineers and the Public*, The Federation Press, Sydney.

Hawken, P. (1993), *The Ecology of Commerce: A Declaration of Sustainability*, HarperBusiness, New York.

Hawken, P., Lovins, A. and Lovins, L. (1999), *Natural Capitalism: The Next Industrial Revolution*, Earthscan, London.

Held, D., McGrew, A., Goldblatt, D. and Perraton, J. (1999), *Global Transformations: Politics, Economics and Culture*, Polity, Cambridge.

Hindmarsh, R. and Lawrence, G. (eds)(2001), *Altered Genes II: The Future*, Scribe, Melbourne.

Ho, M-W. (1998), *Genetic Engineering: Dream or Nightmare*, Third World Network, Penang, Malaysia.

Hoogvelt, A. (1997), *Globalisation and the Postcolonial World: The New Political Economy of Development*, Macmillan, London.

Jones, M. (2001), 'The Contradictions of Globalisation', *Journal of Australian Political Economy*, 48 (December), pp. 5-22.

Kanter, R. (1995), *World Class: Thriving Locally in the Global Economy*, Simon and Schuster, New York.

Kelsey, J. (1995), *Economic Fundamentalism: The New Zealand Experiment - A World Model for Structural Adjustment?*, Pluto Press, London.

Kloppenburg, J. (1991), 'Alternative Agriculture and the New Biotechnologies', *Science as Culture*, 13, pp. 482-505.

Lawrence, G., Higgins, V. and Lockie, S. (eds)(2001), *Environment, Society and Natural Resource Management: Theoretical Perspectives from Australasia and the Americas*, Edward Elgar, Cheltenham, UK.

Legrain, P. (2001), 'The Myth of Global Markets', *The Australian Financial Review*, 10-11 November, pp. 43-44 and p. 48.

Mander, J. and Goldsmith, E. (eds)(1996), *The Case Against the Global Economy*, Sierra Books, San Francisco.

McGrew, A. (19192), 'A Global Society' in S. Hall, D. Held and A. McGrew, (eds) *Modernity and its Futures*, Polity Press, Cambridge (nd).

McMichael, P. (1995), 'The New Colonialism: Global Regulation and the Restructuring of the Interstate System', in D. Smith and J. Borocz (eds), *A New World Order? Global Transformations in the Late Twentieth Century*, Praeger, Connecticut, pp. 37-56.

McMichael, P. (1996), *Development and Social Change: A Global Perspective*, Pine Forge, California.

McMichael, P. (2000a), *Development and Social Change: A Global Perspective*, Second Edition, Pine Forge, California.

McMichael, P. (2000b), 'Global Food Politics', *Monthly Review*, 50(3), July-August, pp. 97-111.

Michelsen, J. (ed.)(2001), 'Politics, Ideology and Practice of Organic Farming', Special Issue, *Sociologia Ruralis* 41(1), January.

Mittelman, J. (ed.)(1996), *Globalisation: Critical Reflections*, Lynne Rienner, Boulder.

Paddon, M. (2001), 'Making Australia Home: Corporations', in C. Sheil (ed.), *Globalisation: Australian Impacts*, UNSW Press, Sydney, pp. 97-126.

Pieterse, J. (1995), 'Globalisation as Hybridization', in M. Featherstone, S. Lash and R. Robertson (eds), *Global Modernities*, Sage, London, pp. 45-68.

Pritchard, B. and McManus, P. (eds)(2000), *Land of Discontent: The Dynamics of Change in Rural and Regional Australia*, UNSW Press, Sydney.

Ritzer, G. (1996), *The McDonaldization of Society*, Second Edition, California, Pine Forge.

Robertson, R. (1995), 'Glocalization: Time-Space and Homogeneity-Heterogeneity', in M. Featherstone, S. Lash and R. Robertson (eds) *Global Modernities*, Sage, London, pp. 25-44.

Ross, R. and Trachte, K. (1990), *Global Capitalism: The New Leviathan*, SUNY Press, Albany.

Sachs, W., Loske, R. and Linz, M. (1998), *Greening of the North: A Post-Industrial Blueprint for Ecology and Equity*, Zed Books, London.

Seitz, J. (1995), *Global Issues: An Introduction*, Blackwell, Massachusetts.

Teeple, G. (1995), *Globalisation and the Decline of Social Reform*, Humanities Press, New Jersey.

Tovey, H. and Blanc, M. (eds)(2001), *Food, Nature and Society: Rural Life in Late Modernity*, Ashgate, Aldershot.

Vanclay, F. and Lawrence, G. (1995), *The Environmental Imperative: Eco-social Concerns for Australian Agriculture*, Rockhampton, CQU Press, Queensland.

Venning, J. and Higgins, J. (eds)(2001), *Towards Sustainability: Emerging Systems for Informing Sustainable Development*, UNSW Press, Sydney.

Walker, K. and Crowley, K. (eds)(1999), *Australian Environmental Policy 2: Studies in Decline and Devolution*, UNSW Press, Sydney.

Waterman, P. (2001), *Globalisation, Social Movements and the New Internationalisms*, Continuum, London.

Waters, M. (1995), *Globalization*, Routledge, London.

Chapter 2

Fast Food/Slow Food: Standardizing and Differentiating Cultures of Food

Mara Miele and Jonathan Murdoch

Introduction

In July 2000, around 30,000 protestors – including greens, communists, anarchists and campaigners from as far away as India and the US – descended on the medieval town of Millau in southwest France. The occasion was the impending trial of Jose Bové, leader of the Confederation Paysanne, a union of small peasant farmers. Bové's crime had been supposedly to 'steal' building materials that were being used in the construction of a local McDonald's. The materials were 'stolen' as part of a protest against import duties imposed by the US Government on a variety of emblematic French goods, including Roquefort cheese for which Millau is famous. According to a McDonald's spokesperson, Bové and the 300 farmers accompanying him incurred damage to the tune of $120,000.

Following his arrest, Bové became a national hero: as one commentator noted, 'the popular media represented him as the last man in France willing to go to jail for the founding ideals of the Republic' (Kingsnorth, 2000, p.38). This was no doubt, in part, because, as Azaryahu (1999, p.485) says, the 'golden arches' have become an icon of global homogenization in culinary tastes, one that is identified with the 'American way of life'. Thus, by attacking McDonald's, Bové was striking out against 'culinary imperialism'. But more than this, he claimed to be standing up 'for the small farmers of France, for traditional methods of production, for the right to be free from corporate hegemony, and for that most French of all causes – gastronomy' (Kingsnorth, 2000, p.39).

While the conflict between the Confederation Paysanne and McDonald's might be seen as simply an expression of national competition between the US and France (in the context of an impending trade war between the NAFTA and the European Union) we believe it refers to something deeper: a widespread and profound unease about the threats posed by the fast-food culture. The unease was expressed by Bové in the following way: 'cooking is culture...all over the world. Every nation, every region, has its own food cultures. Food and farming define people. We cannot let it all go, to be replaced by hamburgers' (quoted in Kingsnorth, 2000, p.40). This sentiment indicates that there are sharply divergent views about the status, significance and function of food in differing cultural contexts. What we see here is the fast food, 'food as fuel' culture of the United States pitted against food as heritage and culture, emblematic of local, regional and national ways of life. Bove's 'action', momentarily at least, threw these cultures of food into stark relief.

In this chapter we wish to investigate the two cultures described briefly above. In doing so we firstly locate our analysis in the context of new food consumption trends in western, 'post-industrial', societies. We begin from the assumption that patterns of food provisioning, purchasing and eating are changing, often in unpredictable ways. The changes, in part, can be ascribed to a context of 'abundance' so that a majority of people living in post-industrial societies have access to a wide variety of foods (frequently offered at cheaper and cheaper prices) (Montanari, 1996).

Secondly, we propose that the two food cultures revealed by Bove can be seen as 'condensed' within particular culinary networks, leading to a recursive relationship between network and cultural context in which each is shaped by the other – see Murdoch and Miele (2002) for an exploration of 'condensation' and 'recursivity'. To illustrate this latter point, we present two case studies: one outlines the development of McDonald's, which is taken as a prime exemplar of a 'fast', 'Americanized' food culture; the second considers Slow Food, a consumer movement that has arisen in order to counteract 'fast culture' (the organization might be seen as a more structured form of resistance than that provided by Bove). These two networks serve to condense cultural 'norms' and in so doing facilitate the spread of particular culinary cultures: McDonald's spreads the suburban values of post war America; Slow Food embodies a European philosophy that 'all food, feeds not only our bodies but also our minds; it is more than just a meal, it is part or our way of life' (Fiddes, 1991, p.6). We assess the spread of these two cultures below.

Even though the two cases represent opposite ends of the food spectrum (one is the archetypal example of standardization, while the other celebrates localized diversity), it is possible that they might coexist and flourish simultaneously. Such co-existence could be seen as illustrative of a 'post-modern' attitude amongst modern consumers in which a range of differing food items are mixed with one another. The existence of this attitude would then lead us to speculate that processes of change in contemporary culinary cultures will not necessarily proceed towards greater homogenization (as the ever-growing number of McDonald's restaurants seems to suggest). Rather, food consumption patterns will become more complex and perhaps more differentiated. As Fabris (1995) says, what we eat is no longer governed by 'unquestioned tradition' but by 'multifarious desires'. Food consumption is, therefore, responsive to changeable market components such as price, taste, lightness, healthiness, convenience, fun, and comfort. Furthermore, the significance of these components varies according to the context of consumption: what is acceptable for a quick lunch, or a snack in between work appointments, would not be appropriate for supper at home with the family and it would be even less acceptable for a special dinner with guests. The varied situations in which food is consumed heightens differentiation in consumption patterns.

The consumption context which interests us here is 'eating out'. As a consequence of food abundance and affluence in North-western societies, there has recently been a sharp decline in home prepared and consumed meals and a corresponding increase in meals consumed outside the home (Gordon, 1998; Warde and Martens, 2000). Fine (1996), in a study of restaurant culture in the US, argues that an increase in the number of restaurants (and cuisine styles) was linked to the number of affluent consumers living in urban areas at the end of World War II:

> The prosperity of post-industrial Western societies, particularly in the last few decades, has provided a fertile breeding ground for new restaurants. This prosperity is both a cause and a result of changes in global markets: with the ability to obtain culinary items from all over the world at all times at prices that consumers can afford, the possibilities of food preparation multiply (Zukin, 1991, p.2 quoted in Fine, 1996, pp.6-7).

This overall trend seems set to continue, with Promar (2000) forecasting that in Western Europe by year 2020 about 40 per cent of all food will be consumed outside the home.

According to Ritzer (1996), in this scenario the process of standardization and rationalization, already emerging in the US food sector with the spread of franchizing theme restaurants (see also Gottdiener, 2000), will increase. However, it should not be assumed that such patterns will sweep through all socio-cultural formations (at local, regional and national levels) in quite the same way. As Ritzer in a more recent work has underlined, the degree to which the McDonaldization process will evolve is going to vary according to the demands of diverse contexts:

As a general rule, restaurants in smaller cities and suburbs will tend to be McDonalidized to a greater degree than those in urban settings. McDonald's began as a smaller city and suburban phenomenon and its strength continues to be in those locales... However, there are a variety of successful models in a city such as New York, many of them with well-established roots and a loyal clientele. Thus, although McDonalidized fast-food restaurants have established a successful presence, other models are likely to survive and continue to prosper (Ritzer and Ovadia, 2000, p.39).

In certain circumstances, then, differentiation in food consumption patterns may prevail. Thus, contemporary food markets hold together a number of divergent tendencies. Some of the most notable of these tendencies are illustrated by our two case networks that have been chosen to represent quite distinct trajectories of development.

Two Culinary Networks

In this section we consider how two global networks have emerged from very different contexts of food preparation and consumption (characterized as cultures of 'food as fuel' and 'food as pleasure'). We illustrate how the two networks 'condense' these consumption cultures. Firstly, we examine McDonald's which, as the Ritzer and Ovadia quote above indicates, serves to define and reflect a set of suburban values and aspirations that were common in the US in the 1950s. Secondly, we consider Slow Food and show it is embedded in the diverse regional cuisines so evident in Italy and other parts of southern Europe. We consider each case in turn before providing a brief comparative analysis.

McDonald's

Suddenly, I felt at home. Because you can take McDonald's out of America, but you can't take America out of McDonald's (Boym, 2001, p.1).

Although McDonald's is the premier example of a globalized network (Mitchell, 2000, p.89), the company can be situated in a particular time and place. It was born in 1950s America, a time of rising wages, a boom in the birth rate, suburbanization, mass ownership of the motorcar, and increased leisure time. As Fine (1996) shows, this was a period when eating out became a standard pastime, with hamburgers and other fast foods increasing in popularity. Hamburgers seemed to capture the 'spirit of the age', its 'capacity for speedy preparation with uniformly satisfactory results...meshing well with...demands of consumer and entrepreneur alike' (Jakle and Sculle, 1999, p.144).

Fast food also sprang from a concern about cost: during this period, despite the growth in demand for their food products, restaurant owners were confronted with inflationary increases in the prices of inputs. They therefore concerned themselves with streamlining and standardizing their operations (Jakle and Sculle, 1999). It was in this context that the McDonald brothers, Maurice and Richard, opened their first restaurant in San Bernardino in California. They too were seeking a way to serve good quality food, at cheap prices. They thus established a 'system' for delivering large quantities of burgers at low cost: 'a 1.6-ounce hamburger, 3.9 inches in diameter, on a 3.5 inch bun with .25 ounces of onion sold for 15 cents – a standardised product of high quality but also low price' (Jakle and Sculle, 1999, p.141; see also Schlosser, 2001). Under their system, customers drove to the restaurant where they parked, walked to a window and placed their orders. The restaurant served few food items, all delivered to a standard format.

The restaurant quickly attracted a great deal of attention, none more so than from a young entrepreneur named Ray Kroc. Kroc was initially attracted to the McDonald's fast food restaurant because of its ability to generate quick profits from a simple system of preparation and consumption. Ritzer (1996) quotes him as saying: 'I was fascinated by the simplicity and effectiveness of the system...each step in producing the limited menu was stripped down to its essence and accomplished with a minimum of effort. They sold hamburgers and cheeseburgers only. The burgers were all fried the same way'. But according to Jakle and Sculle (1999) it was not simply the rationalized food delivery system that appealed to Kroc: he also saw the restaurant as emblematic of a 'suburban lifestyle', one that he himself aspired to. Thus, he entered into negotiations with the McDonald's brothers in order to acquire the franchise. In 1955 Kroc took control of the McDonald's trademark and food delivery system and opened his first restaurant in De Plaines, Illinois, close to the commuter station where he took the train to work. It was his hope that the new McDonald's would draw in a 'youthful, growing, and home-bound trade' (Jakle and Sculle, 1999, p.146) that is, the new suburbanites.

With Kroc in control, the number of restaurants increased rapidly, reaching 200 by 1960 making it the leading fast food chain in the US. Kroc rationalized the McDonald brothers' system even further and ensured that costs were kept to a minimum (Ritzer, 1996). Despite its subsequent growth, the initial suburban aspirations remained at the heart of company policy. Executive Vice-President Fred Turner was quoted as saying: 'We want young families in the tricycle and bicycle neighbourhoods – the station wagon set, or one car going on two' (Ritzer, 1996). In order to attract these suburban families the company concentrated on 'Quality, Service, Cleanliness, and Value' (in Kroc's immortal phrase): good food, for low prices in 'busy, productive controlled, yet joyful' surroundings (Ritzer, 1996, p.148).

The suburban character of McDonald's did however become diluted during its subsequent stages of growth, notably following the company's listing on the US Stock Exchange in 1965. Now profits had to be kept up through a relentless increase in the numbers of restaurants (an increase that continues to this day). During the 1970s and 1980s McDonald's expanded into all major US cities. This move was supplemented by expansion overseas as the company sought to maintain earnings growth. By the mid-1990s it had 25,000 restaurants and operated in almost 120 countries (global earnings at this time stood at around $11 billion). Yet, despite the shift away from suburbia, the company retained aspects of its early character. The Hamburger University – established by Kroc in 1961 to spread good practice throughout the chain and to disseminate a 'family feeling' amongst franchisees – acted to instil core (suburban) values in company employees. The standardized restaurants and products all enshrined the early aspirations for quality food at low prices, served in informal but 'joyful' surroundings. As the company's website puts it the 'McDonald's vision is to be the world's best quick service restaurant experience. Being the best means providing outstanding quality, service, cleanliness and value, so that we make every customer in every restaurant smile'.

Thus, despite its rapid development, McDonald's continues to reflect the suburban living patterns that accompanied the shift from an industrial to a service economy; it celebrates the power of technology in the speed with which food is prepared and served, and, perhaps most important of all, it upholds an informal and easy-going style of food consumption quite at odds with traditional (or 'elitist') modes of dining out (Schlosser, 2001). In these respects, McDonald's has remained true to its roots: it was born in a time and place that celebrated business success, technological prowess, value for money for the consumer, and a standardized but reliable product. These aspects of McDonald's remain intact despite its contemporary global reach.

Clearly, McDonald's has been extremely successful in defining spaces of food consumption in many different places around the world. It has done a great deal to promote the culture of abundance mentioned earlier by cheaply and efficiently distributing burgers and other standardized foods to many millions of consumers in many diverse local settings. The staging of the consumption event by McDonald's has been a triumph of global network building (Murdoch and Miele, 2002). The company seems to have pulled off the trick of disseminating a standardized and rationalized food in a form that permits consumers to experience its products and environments in a variety of different ways (see Finklestein, 1999; Kellner, 1999; Boym, 2001). Boym (2001, p.1) asks:

> What is it about McDonald's that attracts children and immigrants alike? As a rule, immigrants, like children, are very sensitive creatures. In their desire to blend in, they are conscious of making the wrong gesture, looking funny or different, standing out in any conspicuous way. The simple experience of entering a restaurant, asking for a table, and talking to a waiter can be intimidating. In this respect, McDonald's is the ultimate populist place. No one can be excluded, you can come and go as you please. It's okay to bring your children and to make a mess. Toys are given away along with nutritional information: there is something for everyone... The most important populist aspect of McDonald's is, of course, the food'.

Indeed in McDonald's restaurants little knowledge or cultural capital is required to enjoy a meal and this populist aspect is undoubtedly one of the main ingredients of its success.

At the same time the populism of McDonald's has inspired a specific type of criticism or resistance, as in the case of the Hampstead residents in north London who sought to prevent McDonald's opening a restaurant in their up-market high street, or the march in Rome against the opening of a McDonald's restaurant in the famous *Piazza di Spagna*. Beyond this 'elitistic' criticism of McDonald's (which is largely based on the image of the restaurant chain) much more substantive challenges have been made to the spread of McDonald's. For instance, it is accused of producing food that is rich in sugar and salt, a dietary mix that is seen as contributing to obesity (Vidal, 1997). Moreover, it is identified as an agent of standardization in the food chain:

> Behind the front counters of the fast food chains, the familiar menus and logos of McDonald's...lie other assembly-line operations, ownership of which is concentrated in ever fewer hands, allowing even greater economies of scale. Thirteen large slaughter-houses, or meat packing houses in US terminology, now supply most of America's beef. Three companies, Simplot, McCain and Lamb Weston (which is owned by the even larger conglomerate ConAgra), control 80 per cent of the US market for frozen french fries. In the wake of the launch of the Chicken McNugget – made from constituted chicken, flavoured with beef additives and containing twice as much fat for its weight as

hamburgers – eight chicken processors ended up with about two-thirds of the US market (Meek, 2001, p.3; see also Tansey and D'Silva, 1999 and Schlosser, 2001).

In other words, McDonald's is but one part of a profound rationalization of the food sector that extends from plough to plate (Ritzer, 1996). Not surprisingly, this rationalization process has provoked counter-movements, organizations that are dedicated to disseminating a quite different set of food consumption practices. One of the most notable is Slow Food.

Slow Food

> [T]he degree of slowness is directly proportional to the intensity of memory; the degree of speed is directly proportional to the intensity of forgetting (Milan Kundera, *Slowness*).

Slow Food was established in Bra, a small town in the Piedmont region in the North of Italy, in 1986 by a group of food writers and chefs. The immediate motivation was growing concern about the potential impact of McDonald's on food cultures in Italy. The first Italian McDonald's had opened the previous year in Trentino Alto Adige, a region in the North East of Italy. It was quickly followed by a second in Rome. This latter restaurant, because of its location in the famous Piazza di Spagna gave rise to a series of protests (see Resca and Gianola, 1998). These protests provided the spur for the founding of Slow Food.

In the beginning, the movement's founders were concerned that the arrival of McDonald's would threaten not the growing up-market restaurants frequented by the middle/upper class city dwellers, but local *osterie* and *trattorie*, the kinds of places that serve local dishes and which have traditionally been frequented by people of all classes. Because, in the Italian context, traditional eateries retain a close connection to local food production systems, Slow Food argued that their protection required the general promotion of local food cultures. As the Slow Food Manifesto put it, the aim of the movement is to promulgate a new 'philosophy of taste' where the guiding principles should be 'conviviality and the right to taste and pleasure'. Other key objectives include: disseminating and stimulating knowledge of 'material culture' (every product reflects its place of origin and production techniques); preserving the agro-industrial heritage (defending the biodiversity of crops, craft-based food production and traditions); and, protecting the historical, artistic and environmental heritage of traditional foods (the cafés, cake shops, inns, craft workshops and so on) (see http://www.slowfood.com). In short, the movement sought to develop new forms of 'gastronomic associationalism' that link the cultural life of food to biodiverse production spaces.

Slow Food was established on the basis of a local structure, co-ordinated by a central headquarters in Bra. The local branches effectively engage in a range of activities aimed at strengthening local cuisines (see below). These branches were initially established in all the Italian regions (and were called *condotte*) but soon began to spread to other European countries and then further afield (outside Italy the branches are called *convivia*). In 1989 Slow Food was formally launched as an international movement. In that year representatives from twenty countries attended a meeting in Paris and agreed both an international structure and a manifesto. The manifesto asserted 'a firm defence of quiet material pleasure' and stated, 'Our aim is to rediscover the richness and aromas of local cuisines to fight the standardization of Fast Food'. It went on to say: 'our defence should begin at the table... Let us rediscover the flavours and savours of regional cooking and banish the degrading effects of Fast Food... That is what real culture is all about: developing taste rather than demeaning it'. The movement thus began to establish itself outside Italy and at the time of writing convivia exist in 40 countries and the movement has around 70,000 members.

In outlining how Slow Food has developed we describe below the changing character of its activities. In general, the main focus of the movement has been the diffusion of knowledge about typical products and local cuisines to consumers. However, more recently another complementary set of activities has been added. The new activities are aimed at rescuing from 'extinction' the typical products that are facing a dramatic decline in their market (Slow Food, 2000). These are long-term projects that require the co-operation of a large number of actors – farmers, food processors, retailers, local institutions and so forth. We document here the broadening nature of the movement and show it has come to extend the notion of 'gastronomic association' from consumers to producers.

In articulating a response to the spread of McDonald's throughout Italy, Slow Food first began to disseminate information about local food cultures and the challenges they face. In so doing, it effectively became a 'clearing house' for knowledge of local foods, initially in Italy, but latterly more globally. The main means by which knowledge about local and typical cuisines is disseminated is the publishing company, established in 1990. Slow Food Editore publishes a range of guides in order to lead consumers to the food products available in a whole variety of local areas. In the main these refer to Italian cuisines. Thus, alongside the *Vini d'Italia* wine guide, published in collaboration with the *Gambero Rosso* food monthly, Slow

Food publishes *Osterie d'Italia*, a guide to the traditional cuisine of the Italian regions. However, as the movement has internationalized, so its publications have begun to focus upon typical foods found outside Italy. It recently published the *Guida ai Vini del Mondo*, a world wine guide describing as many as 1,900 cellars in 30 countries and *Formaggi d'Europa* a 'fact sheet' on the 127 European DOP cheeses in 1997. The movement's quarterly magazine, *Slow*, is produced in five languages and carries articles on foods from around the world.

The dissemination of knowledge also takes place through the local members. Every Slow Food group is encouraged to organize periodic theme dinners, food and wine tours and tasting courses. The collaboration of the groups underpins national and international initiatives. The following are the most noteworthy: 'Excellentia', a three-day event involving 5,000 people all over Italy in twice-yearly blind tastings of international and Italian wines; 'Taste Week' (*La Settimana del Gusto*), which sets out to familiarise young people with quality catering; 'Friendship Tables' (*Le Tavole Fraterne*), which finance charity initiatives (such as the installation of a canteen in a hospital for Amazonian Indians, the rebuilding of a school in Sarajevo and the restructuring of an Umbrian dairy damaged by the 1997 earthquake); and the 'Hall of Taste', a food fair held every two years in Turin (this is a large, prestigious event that, in 1998, recorded over 120,000 visitors).

In its publications, tastings, talks, conventions and so forth, Slow Food frames regional and local foods in ways that partially isolate them from their surrounding contexts but which retain strong ecological and cultural connections. It readily recognizes that one of the reasons many local and regional food products are disappearing is because they are too embedded in local food cultures and ecologies; they are not easily extracted and sold into modern food markets (there has been little technological or organizational innovation around them, and they often cannot travel the long distances, either for cultural or ecological reasons, covered by McDonalds' burgers). So Slow Food attempts to bring modern consumers to these traditional products and the restaurants in which they are served by stressing their symbolic value. In short, it attempts to bring the products to new markets (or, more accurately, bring new markets to the products).

The activities outlined above are aimed at consumers. However, since it began to identify the importance of local cuisines in maintaining food diversity, Slow Food has also become aware of the problems faced by the producers and processors of the products which compose local cuisines. It has therefore begun to play a more direct role in the protection and promotion of such products. The first of these initiatives was the 'Ark of

Taste', launched in 1996, which aims to 'save from extinction' such typical foods as the tiny production of cured meats (for example, *lardo di colonnata*), artisan cheeses, local varieties of cereals and vegetables, and local breeds (Slow Food, 2000). To assist this activity an Advisory Commission (composed of researchers, journalists and other food 'experts') was formed in order to evaluate products proposed for inclusion in the 'Ark'. The Commission was charged with gathering information on the processing, cultivation or breeding techniques, and commercial potential of the products and also with developing intervention strategies to facilitate their 'rescue'. As part of this project, Slow Food has begun a major 'census' of quality small-scale agro-industrial production and has encouraged Slow Food *osterie* and *trattorie* (those listed in *Osterie d'Italia*) to include the products in their dishes. The Ark project thus aims to enlarge the market for these lesser-known products.

Slow Food is now broadening the range of these producer-oriented initiatives. It recently established local groups ('Praesidia') in order to provide practical assistance to small producers of typical products (organizing commercial workshops, identifying new marketing channels, and similar activities). Another initiative targeted at producers, and aimed at protecting biodiversity, is the 'Slow Food Award'. The first award was given in October 2000 in Bologna (the European City of Culture) to biologists, fishers, and small-scale entrepreneurs whose work helps defend the world's biodiversity. And, in a conscious emulation of McDonald's, Slow Food is about to establish a 'Slow University' which aims to spread good practice in relation to the growing, processing, preparation and consumption of typical products.

This second set of initiatives indicates that the Slow Food movement has entered a new stage in its development. As we have seen, it arose as a response to the arrival of McDonald's in Italy, and claimed to be concerned for local *trattorie* and *osterie*. Since that time, its goals have broadened and the organization has become more complex. Thus, after spending the early years developing the capacity to disseminate knowledge about local cuisines and typical products to middle-class consumers, Slow Food has now started to engage more directly with producers and processors in order to strengthen the local base of typical production. In so doing, it has shifted its attention from the marketing of typical foods to the full range of activities that lie between producer and consumer. In this respect, the movement stands as an example of a sophisticated reaction to the spread of fast food: it extends from the local to global but seeks to put in place sets of gastronomic relations which effectively promote diversity in food as an intrinsic part of cultural and environmental diversity. In this sense Slow Food stands in direct opposition to McDonald's.

'Faking' and 'Making' Localness: Two Routes to Globality

We have described here two culinary networks both of which, to a greater or a lesser extent, have developed global networks in order to disseminate given food consumption practices. These are networks of very different types – McDonald's is a multinational company, Slow Food is a consumer movement – and they promote markedly different food cultures. They represent, as we mentioned above, two contrasting but co-existing trajectories of change within the food sector: one towards the mass-market provision of standardized products, the other towards diversified and dedicated typical foods. They are linked however by the fact that the growth of one (McDonald's) gave rise to the other (Slow Food). Subsequently, both networks have expanded rapidly and have taken differing routes to 'globality'. In this section, we assess these routes.

While McDonald's was born in the suburbs of the US, it now gets most of its profits from sales in other countries (Schlosser, 2001). In the process of becoming 'global', the company has begun to move away from its suburban roots. To begin with, 'globality' has been pursued via a form of cosmopolitanism where global citizens are interrelated by a 'love of McDonald's' (Probyn, 2000, pp.34-5). McDonald's has set itself up as the multinational defender of 'good citizenship and family values', it seeks to 'personify and humanize an interconnected world, where the Big Mac preceded the Internet in bringing us all together' (Probyn, 2000, pp.34-5). Probyn points out that McDonald's uses promotion and advertising campaigns, such as 'It is McTime Now' (which shows happy consumers of Big Macs in different localities at different times), to offer what Soysal (1994) calls 'a postnational model of membership': McDonald's affirms a rhetoric of global citizenship through images of families and community. The advertisements seek to present a community that transcends the traditional sources of identity or sense of belonging like regions and nations, class, ethnicity and age; they propose a new virtual community of 'globalized familial citizens' (Probyn, 2000, p.35). However, 'seemingly the only thing that holds these fragmentary communities together is a common and universal love of McDonald's' (Probyn, 2000, p.43).

However, this cosmopolitan strategy has been accompanied by another: a strategy of 'insiderization'. In an ever increasingly competitive global market McDonald's needs to ensure its acceptance in locations and cultures that are a long way from its suburban base in the US:

> Being truly global means having no particular affiliation to any country or culture, the argument goes. Only then can you truly connect with local markets in what management guru Kenichi Ohmae calls 'insiderization': becoming so embedded in local markets that locals see the company as local... That is why companies like McDonald's, are playing down their American roots, e.g. employing local advertising agencies to deploy local British sense of humour in ads to 'embed' McDonald's in British way of life (Mitchell, 2000, p.91).

As it moves into other cultures so McDonald's 'trims' its image and products to meet the requirements of these cultures. Thus, in Italy McDonald's sells a great deal of pizza while in Scandinavian countries it sells organic milk. It also observes religious sensibilities in Jewish and Muslim cultures. However, this 'insider' strategy can only go so far: McDonald's still exists to disseminate standardized products and practices. Thus, this 'insider' strategy might be termed 'faking localness'.

Slow Food has also embraced cosmopolitanism as it seeks to set the local in the context of the global. Slow Food began as an elite movement of food lovers and food writers based in Italy. It therefore targeted its discursive resources at those well-educated, middle-aged, middle/upper class urban consumers who comprise the main market for local and traditional Italian cuisines. In other words it sought to enrol the new 'omnivorous' consumers (Warde and Martens, 2000), those actively engaged in mastering the complexities of contemporary cuisines, thereby enlarging the market for typical products. In so doing, Slow Food has become part of that cosmopolitan culture that is 'at home in the world', to use Brennan's (1997) phrase. It promotes this culture by giving participants access to food cultures that carry with them, tradition, identity and local complexity.

However, Slow Food has taken the conception of food as embedded in local context one step further by opening up a defence of 'biodiversity'. It has begun to play a more direct role in the protection of local and typical products in the forty countries in which it now operates. The defence of biodiversity started with the Ark project, the rescuing of typical products from extinction, and it continues with opposition to genetically modified foods and support for organic production. According to a leading member of the movement, Slow Food 'is profoundly linked to the values of the land and the past. The preservation of typical products, the protection of species from genetic manipulation, the cultivation of memory and taste education – these are all aspects of this passion of ours for time' (Capatti, 2000). Slow Food now seeks to ensure that food maintains its local connectedness.

In making this shift from a movement concerned primarily with Italian food culture to one oriented to the defence of biodiversity, Slow Food does not aim to insert itself into local cultural formations in order to bend those formations to its own standardized principles and practices (the McDonald's approach), but rather seeks to identify the features that makes those formations distinctive and special. It works with local resources, traditions, cultures and social forms in order to provide a strengthening from 'the bottom up'. As opposed to the local replication of a global form, Slow Food exists as the aggregation of many diverse local cultures and ecologies. It thus concentrates on 'making', rather than 'faking', localness.

The two networks thus combine the local and global in two very different ways: one attempts to make the local a mirror of the global (when you stand in one McDonald's, in a sense, you stand in them all); the other, seeks to sweep up multiple localities into a loosely constructed globality, one that can make local features (cultures, cuisines, ecosystems) transparent and by making them transparent can give them a new value so that any loss (through, for instance, the encroachment of standardized forms) is acknowledged, recorded and, perhaps, resisted. These two differing local/global combinations represent two distinct cultures of food.

Conclusion

We have described here two cultures of food and have examined two networks that act to refine and reproduce these cultures. The first is a culture of standardization and sameness, in which food comes with a reassuring familiarity that is available to all consumers, irrespective of background or affiliation. While this culture still reflects its suburban base, it is spreading via networks such as McDonald's and is bringing more and more parts of the world within its ambit. Yet, as we have noted above, this culture brings many problems in its wake. The diet of hamburgers, french fries and milkshakes is not only alien to many food cultures, but seems to result in ever increasing levels of obesity thereby stimulating a range of health problems (Meek, 2001). Moreover, standardized foods appear to require standardized production practices so that the increased consumption of fast food leads to greater industrialization and concentration in those industries supplying ingredients.

These problems have provoked a backlash amongst many consumers, green activists and intellectuals. We have examined one part of this 'backlash' here: the Slow Food movement. This movement was stimulated to exist by the spread of McDonald's and initially focused its attention on repudiating fast food by extolling the virtues of local and typical products. Slow Food was born in Italy and initially reflected the cultural aspirations that surround local and regional foods in that country. It has subsequently begun to articulate an alternative approach to food; one embedded in the diverse food cultures still to be found in parts of Europe and elsewhere. In these cultures food is part of broader processes of social life. It is also intimately linked to the environment and the existence of diverse ecological conditions. Slow Food now works globally to defend biodiversity.

These two cultures continue in existence: the suburban food cultures of the United States remain the 'heartland' of McDonald's (Ritzer and Ovadia, 2000); the regional food cultures of Italy are still strongly oriented to traditional cuisines. The value of counterposing these two cultures is to show that trajectories of development in the food sector are not in one direction only – towards greater standardization – for differing socio-cultural formations retain differing characteristics. Moreover, culinary networks can act to reinforce or undermine these characteristics. While it is often assumed that the world is progressively being 'McDonaldized' (even Italy!), we have argued that the forces of standardization can provoke counter-movements that serve to strengthen alternative production/ consumption systems. These differing cultures look set to remain as stable features of the culinary landscape.

Yet, drawing sharp distinctions between these cultures may be misleading for it appears that both can be accommodated within modern consumption patterns. Recent studies of consumption behaviour show that consumers are now able to 'mix and match' differing foods and cuisines according to the contexts of consumption. For instance, there is growing evidence that consumers are adopting differing time horizons in their management of food. Thus, on the one hand, the need to 'save time' in food preparation is reflected in a higher demand for prepared foods and higher use of catering services (Warde, 1999; Miele, 2001). In this context, the McDonald's 'fast food' ethos would seem set to prevail. However, on the other hand, the same consumers can also express a more sophisticated demand for 'variety' in food (such as an interest in 'ethnic' cuisines – see Warde and Martens, 2000), and a nostalgia for typical foods (Chaney, 1996), indicating that a 'slower' approach is also possible. Lyon and Colquhoun (1999, p.191) point out that the two cultures co-exist in the following way:

> Many goods and services are sold on the basis of speed, efficiency or explicitly in terms of how much time they can save us. The specific characteristic of speed becomes the master variable on which we distinguish between brands and judge the progress made by manufacturers or providers. Using food as one

example of this phenomenon, we can see the congruity of fast food, microwaveable frozen products, pizza deliveries to your door, street grazing and the fragmentation of family meal occasions. Set against this is an apparent countertendency to imbue the past – when things were slower – with particular symbolic value. Traditional becomes a metaphor for high quality; a description of ingredients and processes that were in more generous measure than could be expected today.

The authors argue that an appreciation 'for nostalgic products and leisure practices is a characteristic feature of urban populations, the one more exposed to the need of efficiently managing time' and 'can be understood not as an irrational, even pathological, reflection of longing for the past but as a coping strategy for the contradictions' (Lyon and Colquhoun, 1999, p.191). Given that such consumers are set to increase in number, we can expect that such forms of 'cross-cultural' consumption will also increase. These considerations imply that the processes of differentiation we have identified here will increase.

We wish to emphasize that we are not proposing that the forces of standardization currently sweeping through the food sector are somehow benign: quite the contrary, and we have attempted to draw attention to some of the pernicious economic, social and environmental side-effects of fast food networks in the course of our analysis. We simply wish to indicate that these forces unfold incrementally, step by step. Despite their seeming ubiquity, they do not (yet) encompass all aspects of food production and consumption. In fact, certain trends at the present time (notably the 'postmodernization' of consumption) hold out the possibility for developing alternative trajectories in the food sector. It is perhaps incumbent upon us to examine these alternatives and to consider whether they have the potential to mount a real challenge to McDonaldization. Slow Food is just one of many possible routes to an alternative globality, an alternative that can strengthen rather than undermine differentiation and diversity in the food sector.

References

Azaryahu, M. (1999), 'McDonald's or Golani Junction? A Case of Contested Place in Israel', *Professional Geographer,* vol. 51, pp. 481-492.

Boym, C. (2001), 'My First McDonald's', *Gastronomica,* vol. 1, pp. 1-6.

Brennan, T. (1997), *At Home in the World: Cosmopolitanism Now,* Harvard University Press, London.

Capatti, A. (2000), 'Traces Left by Time', *Slow,* vol. 17, pp. 4-5.

Chaney, D. (1996), *Lifestyles,* Routledge, London.

Fabris, G. (1995), *Consumatore and Mercato, Le Nuove Regole*, Sperling and Kupfer, Milan.

Fiddes, N. (1991), *Meat*, Routledge, London.

Fine, G. (1996), *Kitchens, The Culture of Restaurant Work*, University of California Press, London.

Finklestein, J. (1999), 'Rich Food: McDonald's and Modern Life', in B. Smart, (ed.), *Resisting McDonaldization*, Sage, London.

Gordon, A. (1998), 'Changes in Food and Drink Consumption, and the Implications for Food Marketing', in *The Future of Food, Long-Term Prospects for the Agro-Food Sector*, OECD Publications Paris.

Gottdiener, M. (ed.) (2000), *New Forms of Consumption, Consumers, Culture, and Commodification*, Rowman and Littlefield Publishers, Lanham.

Jakle, J. and Sculle, K. (1999), *Fast Food: Roadside Restaurants in the Automobile Age*, Johns Hopkins University Press, London.

Kellner, D. (1999), 'Theorizing/Resisting McDonaldization: A Multiperspectivist Approach', in B. Smart, (ed.), *Resisting McDonaldization*, Sage, London.

Kingsnorth, P. (2000), 'Have-a-go Hero', *The Ecologist*, vol. 30, pp. 38-40.

Kundera, M. (1994), *Slowness*, Faber, London.

Lyon, P. and Colquhoun, A. (1999), 'Selectively Living in the Past: Nostalgia and Lifestyle', *Journal of Consumer Studies and Home Economics*, vol. 23(3), pp. 191-196.

Meek, J. (2001), 'We Do Ron Ron, We Do Ron Ron: A Review of "Fast Food Nation" by Eric Schlosser', *London Review of Books*, vol. 23(10), pp. 3-6.

Miele, M. (2001), 'The Changing Passion of Food in Europe', in H. Buller, and K. Hoggart, (eds), *Agricultural Transformation, Food and Environment - Perspectives on European Rural Policy and Planning - Volume 1*, Ashgate, Aldershot.

Mitchell, A. (2000), 'Global Brands or Global Blands?', *Journal of Consumer Studies and Home Economics*, vol. 24(2), pp. 85-93.

Montanari, M. (1996), *The Culture of Food*, Blackwell, Oxford.

Murdoch, J. and Miele, M. (forthcoming 2002), 'Culinary Networks and Cultural Connections: A Conventions Perspective', in A. Hughes and S. Reimer (eds), *The Geography of Commodity Chains*, Pearson Education, London.

Probyn, E. (2000), *Carnal Appetites Food Sex Identities*, Routledge, London and New York.

Promar, (2000), at http://www.produce-studies.com/PROMARUK/.

Resca, M. and Gianola, R. (1998), *McDonald's Una Storia Italiana*, Baldini and Castoldi, Varese.

Ritzer, G. (1996), *The McDonaldization of Society*, Sage, London.

Ritzer, G. and Ovadia, S. (2000), 'The Process of McDonaldization Is Not Uniform, Nor Are Its Settings, Consumers, or the Consumption of Its Goods and Services', in M. Gottdiener (ed.), *New Forms of Consumption, Consumers, Culture, and Commodification*, Rowman and Littlefield Publishers, Lanham.

Schlosser, E. (2001), *Fast Food Nation*, Allen Lane, London.

Slow Food (2000), *The Ark of Taste and the Praesidia*, Bra, Slow Food Editore.

Soysal, Y. (1994), *The Limits of Citizenship*, University of Chicago Press, Chicago.

Tansey, G. and D'Silva, J. (1999), *The Meat Business*, Earthscan, London.

Vidal, J. (1997), *McLibel*, Macmillan, London.

Warde, A. (1999), 'Convenience Food: Space and Timing', *British Food Journal*, vol. 25, pp. 34-48.

Warde, A. (forthcoming), *Eating Globally: Cultural Flows and the Spread Of Ethnic Restaurants* (draft) published by the Department of Sociology, Lancaster University at: http://www.comp.lancaster.ac.uk/sociology.

Warde, A. and Martens, L. (2000), *Eating Out: Social Differentiation, Consumption and Pleasure*, Cambridge University Press, Cambridge.

Chapter 3

Restructuring of the Sugar Supply to the Industrially-Processed Food Chain: The Brazilian Case

Maria Rita Pontes Assumpção

Introduction

Supply chain management has been increasingly recognized as an important factor to be taken into account in company strategies. Provision of products and services for customers, faster and cheaper than those of competitors, motivates managers to engage in joint actions with organizations up- and down- stream from their companies. The objective of this articulation is to create the conditions the company needs to gain a larger value margin for its products, either through reduced costs or through differentiation. This challenge is even more relevant in the processed food chain due to the high percentage of logistics costs in relation to the products' sales prices. Another particularly important challenge in this chain is to create the capacity to continually launch new products to make up for the growing trend towards loss of consumer loyalty to brands.

The volatility of the demand for food products is increased because of constant product innovations, together with new approaches arising from international markets. These innovations are fostered by multinational companies that have established manufacturing units in Brazil, following market liberalization and the stabilization of the Brazilian economy.

This study analyzes the restructuring of sugar supply channels for food production. The analysis is based on the concept of company networks, and how they work, in order to integrate supply operations with each other. The dynamics of the structural transformation of company networks is explained through technological conditioning factors. After a presentation of this theoretical picture, the situation of the Brazilian food market vis-à-vis the globalized economy is discussed. A report of the results of the observations is then produced. The chapter concludes by indicating the three main determining factors of the restructuring process in the sugar industry.

The findings upon which this analysis is based were obtained through multi-case studies considering five pairs of companies (sugar mill and buyer) that engage in the business of supplying sugar as an intermediary product.

Industry Network

Skinner has declared that a company's productive capacity for efficient production and logistic processes is the basis of its corporate strategy. He highlighted the skills related to product quality and cost, speed, and the consistent performance and flexibility of delivery processes and productive systems as the main competitive factors (Skinner, 1969). The literature on production management formalizes the pursuit of these skills with studies on manufacturing strategy, including the dimension of innovation of products and processes (Slack *et al.,* 1997).

Porter (1990) extends Skinner's analysis by proposing the study of a company's chain of values. There is a need, he argues, to understand a company's value activities and how a company relates to its business partners' up- and down- stream activities, for these allow a company to achieve and maintain competitive advantage. Thus, the scope of a company's influence extends beyond its boundaries, with interrelated activities added to the processes of placing its product on the market or of purchasing raw materials. Reflections on the search for a company's competitive advantage associated with the other business processes led to the concept of supply chain management (Cooper *et al.,* 1997).

The basic principle governing supply chain management (SCM) is to ensure greater visibility of the events related to the satisfaction of demand, be it to meet demand by increasing the efficiency of the existing product flow or to respond to expectations regarding new products. Synchronization between the physical flow of products and the flow of information regarding market needs is the greatest challenge for managers. This visibility offers improved conditions for reduced stocks of intermediary products, raw materials, products for industrial processing and end products (Bowersox and Closs, 1996; Christopher, 1997). The result is the speeding up of the supply of requirements through the use of information, as opposed to that of anticipating stocks to meet the demand. Visibility of opportunities for product differentiation feeds the activity of technological management aiding its creation and development.

Faced with the impossibility of controlling end demand, particularly of foods, the companies participating directly or indirectly in the production and distribution of a product organize themselves into networks to absorb the fluctuations between product demand and supply. Closer proximity to the end market is the result when information about what is happening in the consumer environment is shared with companies downstream. The cycle of new product development is reduced by the establishment of relationships with suppliers. This is the concept of tense flow management used by Green and Santos (1991). This concept emerges from the facilities offered by the technology of information and communication (IT) which accelerates data collection and assessment and the production of knowledge. However, these facilities are insufficient for coping with the complexity of managing the different values and interests of the companies that take part in a business network.

Management of the Supply Chain

Slack *et al.*, (1997) present the structure of a company's supply chain as one which comprises all the companies responsible for the physical constitution of the products offered to their buyers, with the exception of those activities that support the flow of the supply chain. These authors are concerned with the development of capacities to meet the goals of operational strategies (that is, cost, quality, consistent production and delivery performance, innovation, and flexibility both in production and in meeting customer orders). They detail ways to manage a company's value activities as defined by Porter, who states that a company's value activities are those that differ physically and technologically from competitors and are necessary to create value for its customers.

Cooper *et al.*, (1997) broaden the scope of analysis for supply chain management, proposing a methodology that considers three interrelated elements: business processes, management components and supply chain structure. Business processes focus on customer service. The components of management are associated with decision-making processes for the actions that take place in the supply chain. The Supply Chain structure is composed of the physical distribution channels of its products as well as the supply channels of its raw materials. The flow of materials and information passes through this structure, which includes the technological sets of the participating companies. The technological resources that support the business processes among companies limit the possibilities of the levels of efficiency and effectiveness of information and physical flows.

The proximity of suppliers can reduce stocks of raw materials for production. Greater frequency of material and component replacement can be achieved by organizing the joint collection of what is needed. The milk-run process establishes the concentration of transport of the materials required by a given buyer. Materials are collected at several suppliers. The dynamics of supply channel restructuring has been oriented towards the concentration of raw materials, with a reduced number of suppliers and/or transactions among companies. In this case there is also co-production, with the anticipation of production phases that had previously been internal at the buyer's level. These facilities exert the function of concentrating components of lower added value than in assembled products. This work reveals changes in this direction that occur in chains of companies with continuous processes.

Higher agility when dealing with, and using, the information, occurs with the application of IT. IT improves the competitive performance of the members of the productive chain as well as of the supply chain and its flows. The distribution of information must ensure communication of substantial content; it must be fast and move from point to point, guaranteeing visibility throughout the productive chain for all its members. These requirements reflect the change in the organizational structure. Instead of hierarchical and sequential decisions handed down through functions, process management requires that information be connected to objects of analysis (buyer, product, sales force, equipment, processes) so as to personalize reports to support group work around the business process. The flow of information facilitated by the IT structure, as observed earlier, is insufficient for the generation and use of knowledge among companies, since it is dependent on the organizational structure that coordinates the joint efforts among those companies.

Changes in the Supply Chain

As pointed out by Cooper and others illustrated above, the three elements considered in supply chain management are interconnected and a change in one of these elements leads to the reorganization of the others. Changes in business processes interfere in the structure of the supply chain. This happens especially if such changes are related to technology. Thus, new deals may also result in new forms of management of the processes interfacing organizations.

The management components maximize the use of the network structure for the success of the business processes. The integration of the three elements can help to achieve the chain's superior performance in relation to that of competitors.

Lundgren reproduces a model, originally presented in Hakansson (1987) in which a network of companies is represented by the composition of two basic sets that express the inter-organizational connections. These are an institutional set (the coordination of exchanges between production and consumption systems) and a technological set (constituted of the production systems, linking resources and activities according to the industrial logics that define the nature of production and consumer activities, knowledge of product technology, production methods and natural resources). Lundgren states that the forms of network coordination, as well as the standards of behaviour among its actors, affect the possibilities for changes in the technological system and vice-versa. Lundgren points to technological innovation as the main factor responsible for changes in the production structure. He sees technological development in company networks as an interactive and cumulative process guided by the search for the solution of problems resulting from joint actions among the companies. The interactions among actors in the network are aimed at inducing and conducting innovations for the improvement of the entire network.

The Productive Sugar Chain and the Supply Network of Industrially Processed Food

The situation of the international sugar market in the 1990s differs greatly from that of the twenty preceding years, when the sugar and alcohol sector was supported by the Brazilian federal programme – the so-called Pro-alcohol programme. It was a market consisting of 90 per cent raw sugar (that is non-processed centrifugalized sugar) and 10 per cent corresponding to centrifugalized white sugar (both refined and non-refined). It was in this market context that substitutes emerged, such as corn isoglucose or HFSC (High Fructose Corn Syrup, a corn syrup with a high fructose content) and other natural and synthetic sweeteners that began to compete in the market with sugarcane- and beet- derived sugar (Veiga Filho, 1998). By the end of the 1980s pressure for change began to occur. Despite continued protectionism in some domestic markets, the liberalization of world trade affected a number of elements including the international sugar market, protectionist policies at home and abroad, the internal market, and the relationships of sugar mills with their buyers, especially those in the industrial market.

There is evidence of intensified competition between the sugar producers of industrialized nations and of developing countries. Cane sugar is losing ground to sweeteners in the industrialized countries. On the other hand, an accelerated growth is observed in the business of economic blocks (NAFTA and the EU market) and among the large regional corporations of the primary processing sectors of agricultural products and the food industry. Such changes have promoted vertical integration between the food industry in developed countries and the agro-industry in emerging countries. There has been a growing focus on the economy of operations (LMC, 1999) guided by advances in biotechnological research.

Szmrecsányi and Moreira (1991) believe there is a need to conceptualize the markets in which the sugar agro-industry operates beyond those in which its main direct product, sugar, is made available. Sugar serves as an intermediary product in technologically more complex processes such as the production of citric acid (by a biotechnological basis) or for composition with other raw materials for supply to the industry of consumer goods.

Processed Food Supply Network

The large multinational conglomerates of the food industry have established Brazilian units for production and distribution. Competitor networks operate within a new pattern characterized by intense flow and by the integrated management of raw materials supply and the physical distribution of products.

Efficiency is even more urgent when one observes that the competition among the local food production units follows the worldwide pattern of competitiveness and must adhere to the technological standards of their international peers. The food producing companies, in contrast, are pressured by the retailers, who have also changed their activities, largely reducing the costs of their operations. The tendency towards the change of forces in the food product supply channel can be seen as retailers have gained the position of market conductors and have strengthened their position in the supply network of the products they offer to the end consumer.

In the face of two facts – intensified competition among the companies of the food industry and the pressure for widespread distribution to increase the operational efficiency of the supply of products to the retail market – the companies in the food industry find themselves forced to reduce the cost of their products, accelerate their customer service cycles and make ongoing product innovations.

A Radical Change in the Concept of Food Product Development: The Complete Base

The diversity of food processes has increased due to the globalization of markets and the constant changes in notions about food. Today the '*Pão de Queijo Mineiro*' is consumed in Japan and the *Kani kawa* already participates in Brazilian consumption. New developments arise from basic research, whether it is biotechnological or from molecular medicine, allowing for continuous innovations in food products.

The value of practicality has also become increasingly important. Given changes in the roles of women in society, fast and easy food preparation, longer conservation times, easy adaptation to different needs and differentiation between ready-to-eat or semi-cooked food have become increasingly important. Moreover, the diversity of new products seeks to meet a variety of conveniences: for babies, for the elderly, for the energetic, and so forth. A new concept has emerged to cope with the growing need for product innovation: that of complete bases or compounds. Compounds are mixtures of ingredients such as juices, aromas, food dyes, acidulants, emulsions, extracts, and additives, among others. They are formulated to present given characteristics that meet the needs of the food and beverage industry. Compounds for the beverage industry may contain vitamins, minerals, energetic ingredients, fibres and others. Liquid and invert sugars (sugars in solution from a chemical reaction) are important means for the incorporation of compounds.

Compounds facilitate the development of new products and the use of compounding contributes towards the acceleration of the processing cycle by reducing the food production time. The modified raw materials – components of the compounds – increase the flexibility, productivity and consistency of the performance of productive processes resulting in the reduction of operational costs, and allowing the food industry network to become more integrated. The stabilization characteristics of some basic components in the ingredient industry, such as yeast, allow for greater control of the various stages of food production. In bread making, other ingredients used in the dough ease its densification, improving the texture of baked products. In the production of cookies, these ingredients avoid the need for batter baking. The use of a modified ingredient mixed with sugar enables uniformity in the formation and production of candies. These modified ingredients also ensure the standardization of different production lots that use seasonal raw materials. Fruit pulp containing these ingredients, for instance, remains fresh for longer periods. Greater process stability and uniformity of component materials and products are ensured.

The compounding industry supplies pre-developed products that are easily applicable when in synergy with the buyer's production plant. The company willing to work with compounding must invest in equipment and its adoption is therefore dependent on a strategic decision. Adoption of compounding must also take into account the market in which this company will act, its distribution, its positioning, the target public to be explored, and the investments in marketing to be targeted at the new product. In addition to product development, compound processors offer their users/customers technical support and other service variables. Supply logistics are simplified by the concentration of several ingredients in a single product, reducing the number of transactions between supplier and buyer.

This industrial segment may substitute for ingredient distributors (aromas, essences, food preservers, and so forth) and raw material suppliers (such as sugar, flour, cocoa) and constitutes the segment of complete bases – compounds – for the food industry. This new element in the food chain has the function of concentrating these raw materials, distributing them to industrial buyers in customized form as new products are developed according to desired specifications.

Within this context, the demand for professionalism in the supply of sugar required by the food industry was far from being achieved, even with the efforts of some sugar mills to develop competences in this field. The issue lies in the capability for technological management of biotechnological processes.

Restructuring of the Sugar Supply Channels to the Food Chain

With global corporations from the food sector entering the Brazilian market, changes have been made in the structure of the competition among them and among companies downstream in the distribution channels.

There have been growing product innovations and – in their local operating units – food-processing corporations must develop the technological capability to adapt their new products to the Brazilian taste. Furthermore, biotechnology makes new modified products available which, when incorporated within the processed food production processes, lead to enhanced uniformity and faster production and extended product lifetime.

The industrial market thus requires sugar to accord with standards and specifications established by the new production processes. From being a direct raw material to the food industry – when its purchase used to be solely based on price – today, through a process of aggregate value, sugar has become an intermediary product forwarded to its buyers through technical sales. Services are added to sugar to ensure its quality and conformity to the specifications of the food recipes in which it is to be used.

The sugar mills look for improved competencies so as to meet specifications demanded by the food industry. This leads to the expansion of their logistic activities and customer services attached to sugar differentiation. This structure also serves the export and retail markets. Most mills sell their sugar to the international markets through trading companies that already act in the international market. Few of them set up their own port terminals, choosing instead to associate themselves with trading companies.

As a new intermediary raw material, sugar as a sweetener changes its structure, becoming liquid and 'invert' sugar. Although the technology has existed for some time, it used to be monopolized by the Dedini group. The Usina São Luiz, which has produced invert sugar since the 1970s, installed two liquid sugar plants in the second half of the 1990s, when the Barra, Nova América, Guarani and Copersucar mills also made technological expansions of their activities. They diversified their activities to differentiate their products, investing in industrial processes to refine and liquefy sugar, producing liquid or invert sugar to obtain syrup composed of glucose, fructose and saccharose. The Nova América mill stands out from the others by seeking the capability of technical sales of its products.

The liquid and invert sugars, by anticipating the syrup activity, involve one phase less – the first one – in the industrial soft drink production process, and one phase less – the last one – in the production of candies, sweets, cookies, juices and crystallized fruit. By offering a ready-for-use product, it simplifies and standardizes the industrial processes involving these products, facilitating their operation and control, besides the elimination of packaging (bags) handling problems. Thus, it offers advantages to its buyers, enabling them to increase their productivity, reduce energy-related losses and labour, and make better use of their space, which was previously used to store sugar bags.

However, the strongest actions towards the diversification of sugar mills in marketing areas, with capability for technical sales of new products for the industrial market, have come about through their association with companies that have been acting more broadly on the market of ingredients for food production since 1998, when the sector went through a serious crisis caused by overproduction.

Dedini Açúcar Líquido started a joint venture with a company with Irish-held capital called Circlet, a member of a group that also acts in the food sector. The liquid and invert sugar producing units were disassociated from the Dedini group and managed separately, supplying invert sugar to the juice, candy, sweets, and dairy industries and liquid sugar to the beverage industry.

Sugar has also become a new intermediary raw material for the food industry through a biotechnological process for the production of acidulants and preservatives. The most direct action observed in this direction is the incorporation of Fermenta (the former Usina de Santa Rosa do Viterbo) by Tate and Lyle with the formation of Mercocitrus. The global Tate and Lyle chain, through its local unit, supplies acidulants to its buyers, who are the global corporations of the food industry that now also have local operating units.

This last transformation was made possible through fermentation biotechnology, in which sugar is transformed into acidulants, substituting products that originated from the chemical industry and which grant improved characteristics to the end products of the food industry in regard to preservation and easy production. In addition to the advantage of being considered 'organic' products, sugar-derived acidulants have stabilising characteristics in industrial processes and preserver characteristics in food products. The transformation of sugar into acidulants had already been done by a conventional process by Fermenta, the former Usina Santa Rosa de Viterbo.

Cargill, in partnership with the Usina Guarani, began to diversify its sugar-related activities after it started operating as a trading company exporting sugar to foreign markets and began to act in the Brazilian and worldwide markets as a supplier of industrial ingredients for the food industry. Cargill also supplies raw materials (organic and chemical fertilizers) for sugarcane production.

Corn Products International (Corn), following the splitting of its food production unit to create BestFoods, has focused its attention on supplying its associated company with ingredients. Looking for partnerships with companies engaged in the primary processing of agricultural products, Corn entered an association with Usina da Barra (Barra), which also acts in the sector of wheat grinding and flour production as well as in other primary agricultural processing sectors. The Corn and Barra alliance ensures the offer of invert sugar to the multinational company, which has exclusivity in its commercialization to the industrial market. Corn Products International diversified its actions in the Brazilian market by entering into an association with a national company of chemical specialties. The interest involved in these associations is the production of compounds. As trading companies, as suppliers of raw materials to the agricultural sector or as buyers of sugar, these companies have financed the sugarcane crop, subordinating the sale of part of the sugar production to financing contracts.

Another way Tate and Lyle operate is through Johnson and Johnson for the distribution of sucralose, a low calorie sugar developed and patented by Tate and Lyle. Sucralose also uses cane or beet sugar as raw materials. Sucralose is a low calorie organic sweetener resistant to high temperatures and therefore suitable for use in industrial production. Sucralose, which serves as a raw material for light products, supplies market niches for products containing lower saccharose content, but of organic origin, by substituting for synthetic sweeteners. Because of their potential risk as carcinogens, the latter have been increasingly rejected for industrial consumption, as a result consumer resistance and/or the partial or total prohibition by sanitary authorities. Usina Albertina uses sucralose, supplied by the global Tate and Lyle network, to produce Sucralight, reselling it as a raw material in the production of light products.

Another change observed in the transformation of the sugar chain is the transport of molasses for the production of lysine (monosodium glutamate) through transporting ducts. The logistic chain of the transport has undergone a paradigmatic change engineered by Aji-No-Moto. This company, of Japanese origin, as a shareholder of UNIVALEM, a sugar mill located in the northeast of São Paulo state, receives its principal raw material – molasses – continuously through the duct, as well as gases for its industrial plant located next to the mill.

Other initiatives to set up infrastructures for logistical changes, such as hydro, rail, road and shipping terminals, have been undertaken by COSAN, Cargill, Tate and Lyle, Copersucar, Nova América and UNIVALEM. Unloading systems have been set up for the products supplied to industrial clients or for shipment of exports.

In the case of liquid and invert sugar supply, the receiving facilities are designed and supervized by the supplier mills. For supplies of molasses through the hydroduct for the production of baking yeast, Aji-No-Moto installed a hydroduct terminal for direct unloading into the tanks at the customer's plant. These installations represent protective barriers against the action of other suppliers, once their exclusive use is established by contract and technical specifications in the links connecting the different systems. These barriers ensure investments in fixed assets specifically for the set-up of the infrastructure for the flow, loading and unloading of products. Linking the production lines of the user customer to the product delivery systems increases the power of the supplier who has the initiative in the undertaking.

This new linking element connecting the sugar chain to the food and beverage industry by means of the production of compounds or through the exclusive use of logistical infrastructures transforms the relationships in sugar sales. The sugar mills now have an intermediary, functional and/or commercial role, to place their products on the industrial market.

Conclusion

The findings of the field survey indicate that the restructuring of the sugar supply channel occurs as follows:

- Incorporation of technical progress in the process of transforming sugar into ingredients for the food industry, under the responsibility of the companies that hold the patents on biotechnologies;
- Holding of the points of transition/transhipment of the physical flow of products up to their delivery to users; and
- Submission of the mills to the established relationships that result in incorporations or alliances with the suppliers of industrial ingredients.

On a tactical level, coordination of the sugar chain now requires capacity for technical sales of liquid and invert sugar. The development of a learning process is necessary between the sugar supplier and the industrial buyer to make the new product, in liquid form, suitable for the industrial processes in which it serves as a raw material.

Another important factor that strengthens the coordination in the hands of global companies downstream from the mills is the submission of the mills to the established relationships that result in formal incorporation or alliances with the suppliers of industrial ingredients. These relationships favour the insertion of global corporations in the Brazilian market. The alignment of managers of local business units with those of a global company facilitates the assimilation of knowledge of the complementary activities that are the object of their diversification. The latter includes technical areas and the management of distribution channels and marketing activities in the local market – including the knowledge that these local units already possess.

Changes in the activities and raw materials of productive and logistic processes, and technical sales of raw materials with added services are, in most cases, interlinked.

This research has brought to light the process that the mills and/or distilleries have undergone to remain in the market. It was believed that these producer units would establish new forms of relations with other industrial segments and that these forms would require technological changes and organizational restructuring that would transform the traditional units into agribusinesses concerned with their corporate performance and also as active participants in the coordination of the sugar chain. However, the coordination of the sugar chain is headed by the producers of industrial ingredients in the 'specialties' segment of the market. This coordination has resulted in the creation of new relations, new activities and productive resources, with the companies operating downstream from the sugar mills controlling the users of these raw materials.

References

Bowersox, D. and Closs, D. (1996), *Logistical Management: The Integrated Supply Chain Process*, McGraw-Hill, New York.

Christopher, M. (1997), *Logística e Gerenciamento da Cadeia de Suprimento: Estratégias para a Redução de Custos e Melhoria dos Serviços*, Pioneira, São Paulo.

Cooper, M., Lambert, D. and Pagh, J. (1997), 'Supply Chain Management: More than a New Name for Logistics', *The International Journal of Logistics Management*, vol. 8(1), pp. 1-13.

Green, R. and Santos, R. (1991), 'Economia de Rede y Reestruturatión del Sector Agroalimentario. Comunicação Apresentada no Seminário: Mudança Técnica y Reestructuración del Sector Agroalimentario', *Conselho Superior de Investigaciones da Espanha*, Madri: 9-11 de Dezembro.

Hakansson, H. (ed.)(1987), *Industrial Technological Development: A Network Approach*, Croom Helm, London.

LMC (1999), 'NAFTA Starch and Fermentation Industry Development', *Starch and Fermentation Analysis*, LMC International.

Lundgren, A. (1995), *Technological Innovation and Network Evolution*, Routledge, New York.

Porter, M. (1990), *Vantagem Competitiva: Criando e Sustentando um Desempenho Superior*, Campus, Rio de Janeiro.

Skinner, W. (1969), 'Manufacturing: Missing Link in Corporate Strategy', *Harvard Business Review*, May-Jun: 156-167.

Slack, N., Chambers, S, Harland, C., Harrison, A and Johnston, R. (1997), *Administração da Produção*, Atlas, São Paulo.

Szmrecsányi, T. and Moreira, E. (1991), 'O desenvolvimento da Agroindústria Canavieira do Brasil Desde a II Guerra Mundial', *Revista de Estudos Avançados*, São Paulo, jan./abr. 11, pp. 59-79.

Veiga Filho, A. (1998), *Mecanização da Colheita da Cana-de-açúcar no Estado de São Paulo: Uma Fronteira de Mecanização Tecnológica da Lavoura*, UNICAMP, Campinas.

Chapter 4

Forging New Local/Global Links Through Fair Trade Agro-Food Networks[1]

Laura T. Raynolds

Introduction

The Fair Trade movement represents an important new challenge to the ecologically and socially destructive relations which characterize the global agro-food system. This movement critiques conventional agricultural production, distribution, and consumption patterns and fosters the re-embedding of international commodity trade in 'more equitable relations'. Fair Trade networks create new multi-faceted market links between progressive consumers in countries of the North and producers pursuing more socially and environmentally sound production in the South. Ultimately, the goal of the Fair Trade movement is to integrate progressive social and environmental principles into conventional international trade practices, thereby creating a more sustainable world agro-food system.

Though the international trade in Fair Trade labelled products is quite young and represents only a minor share of the global market, this trade has grown dramatically in recent years. The world market for Fair Trade products is currently valued at US$400 million, with sales growing at 10 to 25 per cent per year (Fair Trade Federation, 2000). There are currently 800 producer organizations in 45 countries of the South engaged in new Fair Trade networks. Roughly 60 per cent of the Fair Trade market is comprised of food items, such as coffee, bananas, cocoa, sugar, honey, tea, and orange juice (EFTA, 1998). The Fair Trade market is expected to grow rapidly in the near future, fuelled by nascent initiatives in the United States, Canada, and Japan which will make Fair Trade commodities available outside of the movement's home countries in Europe. Planned expansion in the range of commodities eligible for certification under the auspices of the Fair Trade Labelling Organizations International (FLO) will add additional impetus to this budding market. While the Fair Trade movement's market success is impressive, I suggest that its true significance lies not in its market share (which will presumably always be relatively small), but in its ability to create durable new transnational networks which span the widening local/global divide.

Background

Over the past two decades processes of globalization and neo-liberalism have dramatically increased social inequality and ecological destruction around the world. As well as unleashing destructive forces, these processes have given rise to important new social movements organized around the principles of social and ecological sustainability (Beck *et al.*, 1994). Although the ideas and institutions of globalization and neo-liberalism tend to limit the search for alternatives, global free market instability has created important openings for progressive initiatives seeking to re-regulate the world economy (Tabb, 1999). Given the recent erosion in national regulatory capacity, many of these initiatives are oriented toward new local and global political spaces opening up above and below the nation-state (Tickell and Peck, 1995).

People in both the North and the South are challenging conventional global production and trading practices, calling for the tightening of global/local bonds of interdependence. The heightened connectivity between people and places which has made 'globalization from above' possible, has linked resistance movements in the North and South engaged in various efforts at 'globalization from below' (Falk, 1997). As Keck and Sikkink (1998) demonstrate in the case of the human rights, environmental, and women's movements, transnational advocacy networks have effectively integrated marginalized groups into global resource and information systems. Evans (2000) argues that such 'counter-hegemonic' transnational networks can challenge the conventional rules and ideologies of globalization and shift the balance of power in favour of dispossessed groups in the South. Though not focused on the Fair Trade movement, these analyses point to the potential of Fair Trade to challenge exploitative global agro-food principles and practices through the development of networks linking marginalized Southern producers and progressive Northern consumers.

The Fair Trade movement has established multi-faceted transnational networks that provide both an alternative market for goods produced under more socially just and environmentally sound conditions and a basis for challenging destructive conventional world trade practices. The central goal of Fair Trade is to socially re-link producers in the South and consumers in the North, thereby countering the organization of world trade around abstract market principles which devalue and exploit disadvantaged peoples and the environment (Brown, 1993). Fair Trade networks are organized around the multi-directional flow of information, as well as the movement of capital and commodities (Renard, 1999; Whatmore and Thorne, 1997). Northern consumers are informed about the conditions under which items are produced and distributed and encouraged to make positive purchasing decisions. At the same time, producers are informed about Northern market conditions and the activities of other progressive Southern groups and are asked to adopt more socially and environmentally sustainable production practices. In addition to creating new market networks, Fair Trade groups are engaged in direct advocacy to incorporate more socially and environmentally sustainable principles into conventional trade institutions and activities.

While the Fair Trade movement works with handicraft as well as food products, this counter-movement appears to have the greatest potential in challenging our unsustainable global agro-food system. As McMichael (2000, p.21) suggests, the agro-food sector represents a particularly powerful arena for counter-movements, since food links 'nature, human survival, health, culture, and livelihood as a focus of resistance to corporate takeover'. There is increasing recognition that conventional agro-industrial production practices are wreaking havoc on marginal populations and the environment in countries of the South (Murray, 1994; Raynolds, 2000a). In the North, there is growing concern about the safety and social content of food supplies, as evidenced in recent protests against the use of Genetically Modified Organisms (GMOs) in grains, Bovine Growth Hormone (BGH) in milk products, and pesticide residues in fresh fruits and vegetables. Agro-food politics have become a central arena of international debate, increasing the potential future force of consumer-based food movements (see Buttel, 2000).

Fair Trade can be distinguished from a growing array of alternative agro-food market initiatives which promote trade in goods produced under more socially and/or environmentally responsible conditions. As in Fair Trade, initiatives such as organic certification in food and eco-labelling in forest or marine products utilize voluntary labelling systems to assure consumers that certain production conditions have been maintained. Individual labelling initiatives vary greatly – ranging from corporate 'green washing' campaigns to progressive initiatives by non-governmental organizations – but, with the noticeable exception of Fair Trade, they are all concerned only with production conditions.[2] Fair Trade initiatives go beyond the realm of production to acknowledge the conditions under which commodities are exchanged. It is in seeking to establish more equitable trade relations that the Fair Trade movement has developed alternative local/global links as well as alternative products. While the creation of more reciprocal, transnational, agro-food networks represents the most promising aspect of the Fair Trade movement, it also represents the movement's central challenge.

Fair Trade Market Networks

The Fair Trade movement has grown out of a variety of European initiatives seeking to confront 'post-colonialism' through a strategy of 'trade not aid' (Tiffen, 2000). Non-governmental organizations (NGOs) have developed new North/South trade networks that seek to transform market transactions from a vehicle of exploitation to one of producer empowerment. In the words of one of the founding Fair Trade groups, their goal is to 'change international commercial relations in such a way that disadvantaged producers can increase their control over their own future, have a fair and just return for their work, continuity of income and decent working and living conditions through sustainable development' (Fairtrade Foundation, 2000a). Fair Trade items were initially sold in alternative stores. But beginning in 1989, three Fair Trade labels – Max Havelaar, Fairtrade Mark, and TransFair – were introduced in different parts of Europe and facilitated the entry of fairly traded products into conventional sales outlets. The major Fair Trade groups have harmonized their activities and labelling standards under the auspices of the Fairtrade Labelling Organizations International (FLO), which represents 14 European member countries as well as fledgling TransFair organizations in the United States, Canada, and Japan.

FLO and its member organizations are responsible for upholding the integrity of Fair Trade labels through the monitoring of registered producers and the licensing of importers, but are not directly involved in commodity production or trade. One of the aspects which distinguishes Fair Trade initiatives from most certification schemes is that the costs of annual monitoring and certification are paid by Northern importers, not by producers (FAO, 2000). FLO has established Fair Trade certification procedures and requirements for coffee, bananas, cocoa, tea, sugar, honey, orange juice, rice, and wine, using the following core standards:

- Trade must be as direct as possible and based on a long-term commitment by Northern partners;
- Trading partners must pay producers a premium above the world market price and guarantee payment of the FLO minimum price; they must also provide producer credit;
- Producers must be democratically organized associations of small growers or plantations where workers are represented by independent unions;

- Producers must uphold ILO labour standards (including rights to collective bargaining, non-discrimination, minimum social conditions, safe and healthy work conditions, and no child or forced labour); and
- Producers must pursue environmental goals encouraging biodiversity, preventing erosion and water pollution, controlling pesticide and fertilizer use, and reducing wastes.

Where most labelling initiatives involve the certification of a given product, Fair Trade labelling hinges as much on the trade relationship as on production conditions. This is clearly demonstrated in the first two criteria above which require that trade relations be 'direct', 'long-term', and rooted in 'fair, pricing and credit terms'. As Fair Trade labels, brochures, and packages elaborate, connections between Southern producers and Northern 'trade partners' should be based on principles of 'equality' and 'trust'. According to FLO, one of their central goals is 'to set an example of partnership in trade through dialogue, transparency and respect'. The required relationship between North/South Fair Trade partners is multi-directional and multi-faceted, involving not only flows of commodities and money, but also flows of information and expertise. The European Fair Trade Association (EFTA, 1995, p.16) notes:

> The producer-fair trader relationships usually go beyond just selling and buying, and can include the joint development of new products or product lines, the adaptation of products to European fashions, gaining access to new marketing channels, raising investment or working capital and strengthening or expanding the producer organization.

Though Fair Trade labelling criteria deal with the link between Southern producers and Northern imports, multi-faceted Fair Trade networks extend outward to include consumers of labelled products. As TransFair USA (2000) explains:

> Fair trade is an innovative concept that connects producers and consumers in more equitable, more meaningful and more sustainable ways. It happens through the development of consumer interest and the creation/cultivation of socially responsible importers. It is both values and market driven.

As suggested here, Fair Trade networks link consumers, producers, and importers via market and non-market exchanges. While producers draw on these networks to access market expertise and information, consumers use information provided by Fair Trade networks to redevelop trust in the social and environmental origins of their food. As the European Fair Trade Association (1998, p.23) asserts, 'Fair trade "humanizes" the trade process –

making the producer-consumer chain as short as possible so that consumers become aware of the culture, identity, and conditions in which producers live.' Multi-directional and multi-faceted exchanges which go beyond the trade in commodities establish relations of trust between producers and consumers and give substance to Fair Trade networks.

The labelling of products for distribution by supermarkets and institutional suppliers has fuelled the recent expansion of Fair Trade networks in Europe and now in the United States, Canada, and Japan. But with the movement of Fair Trade items into mainstream outlets, labels have become an increasingly important vehicle through which connections are forged between consumers and producers. These labels strive to re-insert social meaning into otherwise abstract commodities, 'speaking to' consumers about issues of authenticity, quality, and naturalness. Fair Trade labels build on the recent rise in 'ethical consumption,' where consumers purchase items which they think have positive, or at least less negative, social implications (Zadek *et al.*, 1998). Fair Trade organizations strive to assure socially-conscious consumers that purchasing products with their label 'guarantees a better deal for Third World producers,' as the Fairtrade Mark sticker reads. The rising market for Fair Trade food products benefits also from burgeoning interest in 'green consumption,' where consumers choose products they believe are less harmful to themselves and to the environment. Fair Trade labels strive to assuage consumers' mounting concern over the origins and content of their food, attesting that items have been produced outside the corporate agro-industrial system responsible for recent outbreaks of food contamination and ongoing global environmental devastation. Many Fair Trade labelled food products are organic, linking the growth of Fair Trade to the global boom in certified organic markets (FAO, 1999a).

Though the market for Fair Trade-labelled products is quite young, annual sales are valued at US$400 million, with recent market growth reaching 30 per cent a year (Fair Trade Federation, 2000). With FLO's ongoing expansion into new commodity areas and the geographic spread of market coverage, Fair Trade sales are likely to grow even more rapidly in upcoming years. To date, there are only a few fairly traded agricultural commodities available – most importantly coffee, bananas, cocoa, and tea – and most of these are only readily available in a handful of European countries. The most well-established product in Europe is Fair Trade coffee, which is sold in 35,000 supermarkets, is served in many municipal, national, and European Union government offices, and now holds three per cent of the entire European coffee market (EFTA, 1998). Total sales of Fair Trade coffee reached over 11,000 metric tons in 1998, with a retail value of over US$200 million (Rice and McLean, 1999, p.78). To date the United States and Canada are not among the major Fair

Trade coffee importers, but dynamic Fair Trade initiatives in both countries are being built around coffee. If successful, these efforts could potentially double world demand for Fair Trade labelled coffee in just a few years (Rice, 2000).

Fair Trade bananas, which were introduced in 1996, have become the second most important Fair Trade commodity in Europe. Though only available in seven countries, an estimated 18,000 metric tons of Fair Trade bananas were sold in 1999 (Max Havelaar, 1999).

In Switzerland where Fair Trade bananas are carried by two major supermarket chains, they have captured almost 14 per cent of the local market. Since their introduction, the market for Fair Trade bananas in Europe has grown at 24 per cent per year (Belling, 2000). Consumer surveys suggest that the total market for Fair Trade Bananas in Europe might be as large as 300-400,000 tons a year (Banana Link, 1997). Fair trade bananas are just beginning to appear in North America; once established they are likely to boost overall demand greatly given the huge size of these markets (FAO, 1999b).

Fair Trade Production

There are currently about 400 producer organizations in 45 countries of the South engaged in the production of the major fairly traded food commodities – coffee, cocoa, tea, and bananas (see Table 4.1). Most of these producers are engaged in the production of coffee, the backbone of the Fair Trade system. Fair Trade coffee is produced by roughly 300 cooperatives representing over 550,000 small-scale growers in 17 countries. Though most producers are in Latin America and the Caribbean, there are also a number of cooperatives located in Africa and Asia (Rice and McLean, 1999, p.58). Fair Trade cocoa is produced by smallholder cooperatives in six countries of Latin America, the Caribbean, and Africa. One of the most successful Fair Trade cocoa cooperatives is in Ghana and involves roughly 30,000 small-scale producers (Swift, 1998).

Though in the early years, Fair Trade networks purchased only from groups of small-scale producers, more recently, producer eligibility has been expanded to include large-scale socially and environmentally progressive enterprises. In seeking to establish partnerships with 'disadvantaged producers', Fair Trade groups have acknowledged that in many countries of the South landless workers are the most seriously disadvantaged. While Fair Trade-labelled coffee and cocoa remains the domain of small-holders, tea and bananas are sourced from both small and large enterprises. There are currently ten grower cooperatives and one plantation producing Fair Trade Bananas in the top five exporting countries. Fair Trade banana producers in the Dominican Republic and new entrants from the Windward Islands are small-holders; in Ecuador and Costa Rica, growers have somewhat larger landholdings. In Ghana a 700-acre plantation, which is known for its progressive management and the strength of its independent union, produces Fair Trade bananas (FLO, 2000). In the Fair Trade tea sector we see a similar range in the types of producers.

Table 4.1   Major fair trade commodities and production characteristics

Commodity	Major Producer Countries	Number of Producer Groups	Type of Producers
Coffee	Tanzania, Uganda, Mexico, Colombia	300	Smallholders
Bananas	Costa Rica, Ghana, Ecuador, Dominican Republic	11	Smallholders and Plantations
Cocoa	Ghana, Bolivia, Belize, Dominican Republic	7	Smallholders
Tea	India, Sri Lanka, Nepal, Tanzania	44	Smallholders and Plantations

Source: FLO internal documents, EFTA (1998)

The requirement that Fair Trade importers pay certification fees, extend producer financing, and commit to long-term purchases, creates more equitable financial and commodity exchanges. Fair Trade networks also open up new avenues of communication, giving producers greater access to market information, technical expertise, and other resources. As suggested by a Fair Trade coffee grower in Peru, 'fair trade is very much in our best interests.

There is a much more direct relationship with them than with most of our buyers' (Ransom, 1995). A coffee grower from Nicaragua adds: 'We have gained a much better knowledge of the international market and of course the price is better...' (Fairtrade Foundation, 2000b). More direct participation in information and communication exchanges within Fair Trade networks gives producers a greater sense of involvement in the international marketing of their produce. A banana grower in Costa Rica describes the importance of his engagement in Fair Trade networks this way: 'Before I was someone that took a box and loaded it onto a train. That was my only responsibility. I was just a farmer...an intermediary. In this new system, I have become an international businessman' (FLO, 2000). Though not all producers are as aware of the nature of the Fair Trade networks in which they are involved as those quoted here (Renard, 1999), studies suggest that the relatively strong North/South linkages between producers, importers, and labelling organizations are what have permitted marginal producers to enter into and benefit from Fair Trade (Blowfield *et al.*, 1999).

As specified in the Fair Trade standards, producers are guaranteed a favourable price. The price for bananas is pegged at US$1.75 per box above the world market price; coffee producers are guaranteed $.05 per pound above the world price (Max Havelaar, 2000). The premium resulting from this pricing system is given to Fair Trade producer organizations to invest in social and environmental activities benefiting their members. The collective investment of this premium can greatly strengthen local groups and provide critical community resources.

Given the extreme volatility of world market prices for tropical exports, for marginal producers the guaranteed price floor for Fair Trade commodities is perhaps as important as the price premium. The importance of the Fair Trade price floor is most clearly demonstrated in coffee, where the world price has fallen below the guaranteed minimum of US$1.26 per pound in seven of the past ten years (Lake and Howe, 1999). When world coffee prices dropped to US$0.50 per pound in the early 1990s, this price guarantee meant the difference between survival and bankruptcy for many small-scale coffee growers. The price floor has also been critical in assuring the survival of marginal banana producers and their families. In the words of an Ecuadorian banana grower, 'Fairtrade is the reason the small farmers in our association still have a livelihood at all' (FLO, 2000). FLO estimates that the Fair Trade Banana circuit transfers roughly one million dollars a year of additional income to banana growers and workers, if we consider the benefits accrued from the Fair Trade price premium and price floor.

Though the Fair Trade pricing system and the active recruitment of disadvantaged producers and progressive plantations appear to provide substantial benefits to those involved, the Fair Trade system is not without problems. Fair Trade networks currently absorb about half of the output of registered Fair Trade groups (Rice and McLean, 1999, p.79). While some see this is as a problem of insufficient demand, others argue that it is important in ensuring that Fair Trade producer groups not become overly dependent on these new markets. Some producer groups have been able to use the information and resources they gain in working with Fair Trade networks to enter other high return markets like that for organic produce (Nigh, 1997), but for some groups this type of market upscaling may not be possible. Producer cooperatives and worker organizations play a critical role in ensuring that potential Fair Trade benefits are fully realized. Even when Fair Trade networks establish meaningful ties between producers, importers, and consumers, Southern producers play a limited role in shaping these networks. Despite Fair Trade's stated commitment to equality, procedures and standards have thus far been developed in the North. To the extent to which these relations reflect the historical dominance of the North over the South, they limit the possibility of creating networks based on 'fairness'.

Conclusion

This analysis suggests that Fair Trade initiatives have begun to create important new multi-directional and multi-faceted networks which span the widening local/global divide. By building transnational networks of solidarity between agro-food producers and consumers, fair trade initiatives encourage the participation and empowerment of disadvantaged Southern producers. These transnational networks thwart the entrance of transnational corporations which increasingly target growing markets for certified products, appropriating the value added by new labels without adhering to the underlying social and environmental values. The evidence reviewed here suggests that Fair Trade can indeed create alternative market networks which bypass our conventional corporate dominated world agro-food system.

The greatest challenge for Fair Trade initiatives may be in ensuring that they go beyond creating a parallel (perhaps necessarily ancillary) market for alternative agro-food products in the North, to integrate more sustainable environmental and social practices into conventional world trade. Fair Trade demonstrates that more socially just and environmentally sound production in the South is possible and that Northern consumers will pay more for these products, thus countering the tendency to see conventional agro-food production and trade practices as inevitable. But can this movement really hope to transform our exploitative global agro-food system? To take on this broader challenge, Fair Trade must go beyond the realm of consumer politics, where individuals with discretionary income make positive purchasing decisions, to the realm of citizen politics, where people make positive collective decisions about the nature of acceptable production and trade practices. The real challenge will be to see if the Fair Trade movement can in fact challenge the conventional rules and ideologies of conventional trade and shift the historical North/South imbalance in world trade.

Notes

1 This work has benefited from the support of the John D. and Catherine T. MacArthur Foundation's Program on Global Security and Sustainability and from collaborative research with Douglas Murray. The views presented here are the responsibility of the author alone.
2 These initiatives vary in their sectoral bases, areas of primary concern, strategies for bringing about change, and non-governmental organization, industry, and government involvement (see Blowfield 1999; Dudley *et al.*, 1997; FAO, 2000; Raynolds, 2000b).

References

Banana Link (1997), 'Massive Potential Consumer Support for EU Fair Trade Bananas', *Banana Trade News Bulletin*, November.

Beck, U., Giddens, A., and Lash, S. (1994), *Reflexive Modernization*, Stanford University Press, California.

Belling, R. (2000), 'Fair Trade in the North: Achievements and Challenges', Paper presented at Colorado State University Fair Trade Research Conference, Keystone CO, May 18-20.

Blowfield, M. (1999), 'Ethical Trade: A Review of Developments and Issues', *Third World Quarterly,* vol. 20, pp. 753-770.

Blowfield, M., Malins, A., Maynard, B. and Nelson, V. (1999), *Ethical Trade and Sustainable Rural Livelihoods*, Natural Resources and Ethical Trade Programme, Kent, UK.

Brown, M. (1993), *Fair Trade*, Zed Press, London.

Buttel, F. (2000), 'The Recombinant BGH Controversy in the United States: Toward a New Consumption Politics of Food?', *Agriculture and Human Values,* vol. 17, pp. 5-20.

Chambron, A. and Smith, A. (1998), 'Bananas: Paradise or Jungle?', in EFTA (ed.) *Fair Trade Yearbook: Towards 2000,* Druk in de Weer, Belgium, pp. 83-101.

Dudley, N., Elliot, C., and Stolton, S. (1997), 'A Framework for Environmental Labelling', *Environment,* vol. 39, pp. 16-20; 42-45.

EFTA (European Fair Trade Association) (1995), *Fair Trade Yearbook 1995,* EFTA, Druk in de Weer, Belgium.

EFTA (1998), *Fair Trade Yearbook: Towards 2000*, EFTA, Druk in de Weer, Belgium.

Evans, P. (2000) 'Fighting Marginalization with Transnational Networks: Counter Hegemonic Globalisation', *Contemporary Sociology,* vol. 29, pp. 230-41.

Fair Trade Federation (2000), *Fair Trade Facts* http://ww.fairtradefederation.com/ab_facts.html.

Fairtrade Foundation (2000a), *The Fairtrade Foundation*, http://www.fairtrade.org.uk.

Fairtrade Foundation (2000b), *An Interview with Juanita Garcia*, http://www.gn.apc.org/fairtrade/nicaragu.htm.

Falk, R. (1997), 'Resisting "Globalisation-from-above" Through "Globalisation-from-below"', *New Political Economy*, vol. 2, pp. 17-24.

FAO (United Nations Food and Agriculture Organization) (1999a), *Organic Agriculture, COAG/99/9*, FAO, Rome.

FAO (1999b), *The Market for "Organic" and "Fair-Trade Bananas"*, CCP:BA/TF 99/7, FAO, Rome.

FAO (2000), *Ad-hoc Expert Meeting on Socially and Environmentally Responsible Banana Production and Trade*, FAO, Rome.

FLO (Fairtrade Labelling Organizations International) (2000), *The Impact of Fairtrade Bananas*, FLO, Bonn.

Keck, M. and Sikkink, K. (1998), *Activists Beyond Borders: Advocacy Networks in International Politics*, Cornell University Press, Ithaca.

Lake, R. and Howe, C. (1999), *The Development Impact of Fair Trade: Evidence from the Work of Tradecraft and Challenges for the Future*, http://www/traidcraft.co.uk/impact.htm.

Max Havelaar (1999), *Fair Trade Banana Statistics*, Max Havelaar, Denmark.

Max Havelaar (2000), http://www.maxhavelaar.dk.

McMichael, P. (2000), 'The Power of Food', *Agriculture and Human Values*, vol. 17, pp. 21-33.

Murray, D. (1994), *Cultivating Crisis: The Human Cost of Pesticides in Latin America*, University of Texas Press, Austin.

Nigh, R. (1997), 'Organic Agriculture and Globalisation: A Maya Associative Corporation in Chiapas, Mexico', *Human Organization*, vol. 56(4), pp. 427-36.

Ransom, D. (1995), 'What's Brewing', *New Internationalist*, p. 271.

Raynolds, L. (2000a), 'Globalisation, Restructuring, and Gender in Caribbean Agriculture', *Global Development Studies*, vol. 2(1-2), pp. 175-98.

Raynolds, L. (2000b), 'Re-Embedding Global Agriculture: The International Organic and Fair Trade Movements', *Agriculture and Human Values*, vol. 17, pp. 297-309.

Renard, M.C. (1999), 'The Interstices of Globalisation: The Example of Fair Coffee', *Sociologia Ruralis*, vol. 39, pp. 484-500.

Rice, P. (2000), *Fair Trade in the North: The United States Experience*, Paper presented at the Colorado State University Fair Trade Research Conference, Keystone, CO, May 18-20.

Rice, P. and McLean, P. (1999), *Sustainable Coffee at the Crossroads*, Paper prepared for The Consumer's Choice Council, Washington, DC.

Swift, R. (1998), 'Best of the Best', *New Internationalist*, August, pp. 20-21.

Tabb, W. (1999), 'Progressive Globalism: Challenging the Audacity of Capital', *Monthly Review*, vol. 50, pp. 1-10.

Tickell, A. and Peck, J. (1995), 'Social Regulation after Fordism: Regulation Theory, Neo-liberalism and the Global-local Nexus', *Economy and Society*, vol. 24, pp. 357-86.

Tiffen, P. (2000), *Fair Trade Production: Dilemmas in the South*, Paper presented at the Colorado State University Fair Trade Research Conference, Keystone, CO, May 18-20.

TransFair USA. (2000), *TransFair USA*, http://www.transfairusa.org.

Whatmore, S. and Thorne. L. (1997), 'Nourishing Networks: Alternative Geographies of Food'. in D. Goodman, and M. Watts (eds), *Globalising Food: Agrarian Questions and Global Restructuring*, Routledge, New York, pp. 287-304.

Zadek, S., Lingayah, S., and Forstater, M. (1998), *Social Labels: Tools for Ethical Trade – Executive Summary*, Office for Official Publications of the European Communities, Luxemburg.

Chapter 5

The Power of Food

Philip McMichael

Introduction

Food poses an interesting paradox at the beginning of the Twenty-first Century. Long discounted by the industrial fixation of development theorists and planners, food (and its security) looms as a force that threatens the current hegemony of the market. In fact, one might venture to suggest that food is as much a force to be reckoned with, as is money. Just as international monetary relations today are fragile, requiring continual ad hoc adjustments, and countered by the expansion of alternative currency movements, so the world food order is increasingly fragile, supplemented by ad hoc food assistance programs, and countered by alternative agricultures.

In order to explore the food question, it is necessary to locate its parameters. I propose to address this by situating it in the crisis of development. There are two parts to this. First, 'development' was synonymous with industrialization – a movement signified in Britain's rise to power as workshop of the world. The ensuing competitive dynamic in the capitalist world was industrial rivalry, extending to the United States, and indeed to the former Soviet Union. As such, development became associated with industrial rationality. It viewed nature as an unproblematic human laboratory and rendered rural society as a residual domain – supplying labour for urban-industrial ventures as agro-industrialization expelled rural populations from their local agricultural communities. In this movement, food was removed from its direct link to local ecology and culture, and became an input in urban diets and industrial processing plants. While this movement continues today, there is a counter-movement towards community agriculture and fresh and organic food that corresponds to the excesses of industrialism and the crisis of development.

Second, the crisis of development refers to the breakdown of consensus regarding the feasibility and credibility of 'development'. Gilbert Rist has suggested development is 'like a dead star whose light can still be seen, even though it went out for ever long ago' (1997, p.230). The crisis of development, as a national project, has been displaced to the global arena, and further debased. Through the enlistment of multilateral institutions, and the pressure to maintain currency credibility in a global money market governed by speculation and securitization (credit management), national governments are busy co-authoring the rules of a global market order, arguably with false, or even disingenuous, hopes that membership will bring prosperity to their countries. This hasty and short-sighted pursuit of globalization expresses the crisis of development.

Globalization is not, however, the only expression of this crisis. There is a plethora of alternatives – including community supported and sustainable agriculture, community food security coalitions, organic food, principles of biodiversity, vegetarianism, fair trade movements, and eco-feminism, for instance. These counter-movements also constitute the crisis of development, offering alternative solutions and trajectories to the globalist response. The fundamental difference between these two responses is often understood in terms of a global/local distinction – between the abstracted perspective (and abstracting force) of globalization, and the particularizing perspective and force of counter-movements. Let me elaborate upon these two opposing expressions of the crisis of development.

The Global Corporate Regime

Whether, and to what extent, a corporate regime comes to dominate world food systems will depend on its political sustainability. By framing the question this way, I emphasize that not only is globalization profoundly political, but also the agribusiness project is itself open to continual modification from the constraints imposed by the natural environment as well as the social counter-movements.

The corporate regime is a set of power relations where formal rules and operating procedures are subject to continual contention – and resistance comes not only from the counter-movements, but the agents of the regime itself. Thus, while there is a broad political counter-movement to the WTO institutional order (expressed in the failure thus far to implement a profoundly undemocratic Multilateral Agreement on Investment that seeks to establish a charter of rights for capital at the expense of citizen rights), the current trade war between the European Union and the United States, beginning with the issue of European banana imports from Latin America and extending to genetically-modified agricultural products and food exports from the US, has triggered a *general* trade war between the US and the EU. In other words, the corporate regime, as institutionalized in the World Trade Organization (WTO), is incomplete and contradictory – precisely because the world order is authored by competing and unequal nation-states, some of which view multilateral rules as sources of national corporate power.

I would not want to suggest that reality has rendered the globalist vision ineffectual. Quite the contrary, the global reorganization of food cultures is extensive and has irreversible social and environmental impacts. Furthermore, private-controlled biotechnology threatens to radically intensify these impacts. It is important to understand the *modus operandi* of the corporate sector.

Operating Principles of the Global Corporate Regime

There are strong reasons to suggest that globalization is a higher-order version of the development project (see McMichael, 2000; McMichael and Lawrence, 2001). Instead of the initial mid-Twentieth Century representation of development as nationally-organized industrial growth, development has now been reframed as globally managed growth, with information technologies and biotechnologies as the leading sectors. In addition to the shift from industrial to post-industrial technologies, a far-reaching shift in political governance is in the works: the elevation of market rule. Whereas development implied a national public sector to regulate the market and its outcomes, globalism seeks to discard or weaken the public welfare function in order to elevate the logic of the market.

That does not mean states are disappearing: rather, they are modifying their role. Markets are political institutions. Whereas in the Nineteenth Century states constructed markets, today states/multilateral institutions are reconstructing markets by restructuring states and their inter-relations.

Under the dictates of debt rescheduling and market reforms, rural regions across the South are being reconstructed as agro-export platforms. Local food security is compromised by the appropriation of land for the fruits of the 'second green revolution', and by the use of dumping and other concessions to install the globalizers' version of food security: food dependence on world 'breadbasket' regions. In fact, half of the foreign exchange of the FAO's 88 low-income food deficit countries goes to food imports (LeQuesne, 1997).

The current restructuring of world agriculture intensifies a global division of agriculture labour, where trade in low-value temperate cereals and oilseeds has been historically dominated by the North, and trade in high-value products has distributed increasingly to corporate agro-exporters (or their contract farmers) producing in the South. For Southern states this is often an unstable trade, signalling a more fundamental process at work: a widespread subordination of producing regions to global production and consumption relations organized by transnational food companies.

Under these conditions, which affect certain world regions, agriculture becomes less and less an anchor of societies, states and cultures, and more and more a tenuous component of corporate global sourcing strategies. It increasingly anchors a system of global profiteering in food products, a system in which food travels from farm gate to dinner plate at an average of two thousand miles. Transnational corporations stand to gain overall from a free trade regime, since it would enhance and reward capital mobility and facilitate it by reducing institutionalized costs.

Global firms have been key supporters of the General Agreement on Tariffs and Trade (GATT) multilateral approach to liberalization. In fact, the original US proposal to the Uruguay Round was drafted by Cargill's former senior vice president, also a former officer of the US Department of Agriculture. Cargill shares roughly 50 per cent of US grain exports with Continental. Food companies, grain traders and the chemical industry all generally favour using the WTO to phase out farm programs, eliminating supply management and driving down prices by exposing producers to world-wide differential labour costs. By reducing price supports, the corporations maximize their ability to structure comparative advantages in the world market, and to source their inputs from the variety of producing regions incorporated into the 'free' world market.

The GATT agreement challenges agricultural supply management boards on the grounds that they interfere with the free trade of agricultural products on the world market. For instance, in Canada supply-management agencies that emerged in the Twentieth Century to protect farmers from corporate food processors, were subsequently challenged by Cargill Canada (Kneen, 1995, p.10). Companies like Cargill have supported the NAFTA and GATT in an attempt to institutionalize a trade regime outlawing such 'distortions' to global markets, despite sustained national and international protests by farmer organizations.

The corporate assault on national regulatory policies is both a trading, as well as a production, strategy. Companies are looking either to capture new markets through the purchasing of crops and processed food, or to directly organize agricultural production. New forms of mass marketing of commodities produced under contract in multiple locations are emerging, especially in the global fruit and vegetable industry. The global coordination of multiple production sites, for a year-round supply of fresh produce, is achieved through information technologies. In Chile, now the largest supplier of off-season fruits and vegetables to Europe and North America, more than 50 per cent of fruit exports are controlled by five TNCs (Watkins, 1996, p.251).

Tomatoes and the WTO

Deborah Barndt's research on the tomato is telling here. With her research team she retraces the journey of the tomato from Mexico to the ubiquitous McDonald's outlets in North America. Naming it 'Tomasita' to foreground its labour origins in national and gendered terms, Barndt describes the Sayula plant of one of Mexico's largest agro-exporters: Santa Anita Packers:

> At peak season the plant employs over two thousand pickers and seven hundred packers. The seeds it uses are hybrid seeds which, although originating in Mexico, were developed and patented outside of Mexico, mostly coming from Israel or the United States. The field production requires the heavy use of pesticides, in multiple varieties. During our visit to the Sayula operation we saw these pesticides stacked in storage, mixed in enormous vats, and being sprayed by trucks that looked like giant anteaters. Many (but not all) field workers covered their mouths with towels and some wore gloves to protect themselves from direct contact with the pesticides; the company did not provide any health and safety education or protective gear. There were reports of children from a school located next to the fields being hospitalized the year before for intoxication by fumes that spread into the school yard (1997, pp.59-61).

Although the impact of these practices on the health of the land was rarely noted, an exposé of the tomato industry in a state paper did disclose that the same companies had left another part of the state four years earlier because the land there had been wasted.

Perhaps a more visually striking indicator of monocultural production was the packing plant, employing hundreds of young women whom the company moved by season from one site to another as a kind of:

> 'mobile maquiladora'...the only Mexican inputs are the land, the sun, and the workers... The South has been the source of the seeds, while the North has the biotechnology to alter them. An agreement such as the 1993 Convention on Biotechnology, while supporting national sovereignty, legalizes companies' (mainly Northern) ownership over *improved* biomaterials and does not really protect farmers' or governments' rights to their biomaterials...the workers who produce the tomatoes do not benefit. Their role in agro-export production also

denies them participation in subsistence agriculture, especially since the peso crisis in 1995, which has forced migrant workers to move to even more scattered work sites. They now travel most of the year – with little time to grow food on their own plots in their home communities (Barndt, 1997, pp.59-61).

Barndt concludes:

> with this loss of control comes a spiritual loss, and a loss of a knowledge of seeds, of organic fertilizers and pesticides, of sustainable practices such as crop rotation or leaving the land fallow for a year – practices that had maintained the land for millennia (1997, pp.61-62).

The WTO is designed to operate as an enforcement mechanism of market rules for the dominant global states and corporations. It is also a tribunal for enforcing corporate rights to manage consumption. On the horizon is an intensification of agrochemical corporate domination of world food production by five conglomerates involved in genetically-engineered food (AstraZeneca, DuPont, Monsanto/Pharmacia, Novartis and Aventis). They account for 60 per cent of the global pesticide market, 23 per cent of the global seed market and almost 100 per cent of the transgenic seed market (Gorelick 2000, p.30). Their sponsorship of upwards of 900 hectares of genetically-engineered crops, portends a controversial future. The companies argue that these new biotechnologies reduce the use of pesticides, and promise an end to world hunger. Critics dispute these claims, arguing also that these technologies will discriminate against small farmers, threaten public health and environmental health and narrow available food choices (see Hindmarsh and Lawrence, 2001).

Bill Heffernan's Report to the US National Farmers Union, entitled *Consolidation in the Food and Agriculture System,* identifies an emerging process of centralization of agro-food capitals involving food chain clusters of 'firms that control the food system from gene to supermarket shelf' Heffernan notes that since the corporate world is so fluid, other acquisitions are necessary to survive. Hence, when Cargill acquired Continental Grain, it meant that Cargill 'would control more than 40 per cent of all US corn exports, a third of all soybeans exports and at least 20 per cent of wheat exports' (Grainnet, cited in Heffernan, 1999, p.6).

Another cluster is Novartis/ADM. Novartis formed through a Swiss merger of CIBA-Geigy and Sandoz, with agribusiness operations in 50 countries, focusing on crop protection chemicals, seeds and animal health. This merger, followed by the acquisition of Merck made Novartis the leading agrochemical firm, with 15 per cent of the global agrochemical market. In the late 1990s Novartis entered into collaboration with the

University of California-Berkeley, agreeing to work 'in all areas of functional genomics related to agriculture, including gene-library construction, sequencing, mapping and bio-informatics' (Heffernan, 1999, p.8). Novartis' genes, seeds and chemicals complement ADM's global grain collection and processing network.

Agriculture constitutes about 65 per cent of the global economy, and corporate centralization is unsurprising. In 1998, Jeremy Rifkin noted:

> The top ten agrochemical companies control 81 per cent of the $29 billion global agrochemical market. Ten life science companies control 37 per cent of the $15 billion per year global seed market. The world's ten major pharmaceutical companies control 47 per cent of the $197 billion pharmaceutical market. Ten global firms now control 43 per cent of the $15 billion veterinary pharmaceutical trade (1998, p.68).

Also ten transnational food and beverage companies' combined sales exceeded $211 billion in 1995 (1998, p.68). Such corporate clustering is complemented by lobbying to revise world food safety standards in favour of genetically-engineered foods, food disparagement laws gaining ground in the US, global PR firms structuring debate in favour of genetic engineering, and, finally, the WTO being deployed to challenge governments that oppose genetically-engineered crops. For example, in September 1997, the WTO ruled against the EU's ban on imported beef and milk from cattle treated with Monsanto's growth hormone, Posilac.

Behind the apparent multilateralism of the WTO stands the attempt to institutionalize rules of a neo-liberal world order to match (and deepen) the corporate led economic integration underway. A more far-reaching, substantive, power is anticipated in the negotiation over the terms of the WTO. In particular, the current dispute over the reach of the WTO regarding investment concerns the institutionalization of a global property regime. Through the TRIPs protocol, trade-related intellectual property rights of foreign investors have been strengthened by the possibility of patenting a variety of products and processes. Global corporations are empowered by this protocol, for example, to patent genetic materials such as seed germplasm, potentially endangering the rights of farmers to plant their crops on the grounds of patent infringement. This is an extraordinary form of expropriation of genetic resources developed by peasants, forest dwellers and local communities over centuries of cultural experimentation.

The WTO is a vehicle for the installation of a corporate regime, targeting food self-sufficiency and food safety as restraints on the market and private accumulation; and perhaps most significantly, seeking to *control the institutional dimension of the world market* – namely the regulatory framework at the international and, by extension, the national, levels, since states author and abide by multilateral rules (Greenfield, 1999). Stephen Gill has termed this the 'new constitutionalism', the removal of political decision-making away from democratic polities and embedding decisions in remote, confidential, and bureaucratic organizations. Through authorship of multilateral agreements, states are embedded in the world market, becoming corporate entities themselves (McMichael, 1995; Greenfield, 1999).

The scenario of a fully globalized food system is undoubtedly far fetched. Roughly 90 per cent of the world's food consumption occurs in the country in which it is produced. Sixty per cent of the food consumed by rural populations they produce, whereas urbanites depend on the market for 90 per cent of their food consumption (McCalla, 1999, p.3). Only about one-fifth of the world's almost 6 billion people participate in the cash or consumer credit economy. Nevertheless, the reality is that full implementation of a WTO regime would be tremendously destabilizing to the remaining three billion people who live from the land, would intensify environmental jeopardy (especially with the growing threat of biological, rather than simply chemical, pollution), and would continue the process of reducing global biodiversity to agro-industrial monocultures.

Justification for this scenario depends on five principal discursive claims of the corporate regime:

Biotechnology's Potential for Feeding an Increasingly Hungry, or Food-Deficient, World

Monsanto corporation's home page has proclaimed: 'Guess Who's Coming to Dinner? 10 billion by 2030'. It warns us that low-tech agriculture 'will not produce sufficient crop yield increases and improvements to feed the world's burgeoning population' declaring that 'biotechnology innovations will triple crop yields without requiring any additional farmland, saving valuable rainforests and animal habitats' and that 'biotechnology can feed the world...let the harvest begin' (Kimbrell, 1998, p.294). The point here is not to debate corporate claims regarding the potential of genetic engineering in increasing food supplies – 'in theory, it is possible that some transgenic plants could be more nutritious, travel better, or produce better yields in harsh climates' (Bruno, 1998, p.293). Rather, the issue is explaining the conditions contributing to hunger. Outside Africa, gains in food production since 1950 have exceeded population growth everywhere, and yet Africa has 'enormous still unexploited potential to grow food' with land to give and grain yields potentially higher than in the North. And 78 per cent of all malnourished children under five in the South live in countries with food surpluses (Lappé and Bailey, 1998, pp.8-11).

The issue is distributional and organizational. Not only do half a billion rural people lack access to land to grow their own food largely because of the agro-export bias, but also, and related, a substantial proportion of commercial food production supplies affluent diets that are unsustainable (see Lappé and Bailey, 1998).

Sustainable Agriculture

Monsanto's CEO Robert Shapiro was quoted as saying that 'sustainable development will be a primary emphasis in everything we do' (see Bruno, 1998, p.292). Arguably, the embrace of transgenic technology seriously threatens not only sustainable development, but also sustainability in general. The issue here is the replacement of diversity by monopoly. In describing the 'commercial enclosure of the world's seeds' Rifkin notes that hundreds of millions of farmers across the world controlled their seed stocks, and their reproduction, just a century ago, whereas today 'much of the seed stock has been bought up, engineered, and patented by global companies and kept in the form of intellectual property' where farmers become simply a new market for genetically-altered seeds (1998, p.114). Further, the large-scale introduction of transgenic crops could contaminate remaining centres of crop diversity through gene drift from transgenic plants to landraces – gene-complexes with multiple forms of resistance to disease.

Efficient Agriculture

Breeding crops for resistance to herbicides may improve yields, but there is a fuzzy logic here. The likelihood of intensified use of herbicides would further attack biodiversity and undermine rural survival strategies. In many rural cultures, non-crop plants (often termed weeds) represent food, fodder and medicine. For example, peasant women in India use 150 different species of plants for vegetables, fodder and health-care; and in Veracruz, Mexico, peasants make use of 435 wild plant and animal species (Mendelson, 1998, p.272).

Getting Government out of (the Food) Business

In addition to reducing international food safety standards, governments are writing themselves out of the picture to the extent that they codify interstate trade relations, from global multilateral, to regional free trade, agreements. Free trade agreements like NAFTA, and the forthcoming FTAA, mirror the asymmetry of the WTO regime. For example, quotas on duty free US corn, wheat and rice imports into Mexico are being lowered in stages. In Mexico, two and a half million households engage in rainfed maize production, with a productivity differential of two-to-three tons per hectare compared with seven and a half tons per hectare in the American mid-West. With an estimate of a 200 per cent rise in corn imports under NAFTA's full implementation by 2008, it is expected that more than two-thirds of Mexican corn production will not survive the competition (Watkins, 1996, p.251). Meanwhile, it is no secret to Southern states that Northern states such as the EU and the US continue to indirectly subsidize export agricultures. In 1995, the farm subsidy bill in the North was collectively $182 billion – 41 per cent of the value of production. Watkins has estimated that the average subsidy to US corn farmers and grain traders is 100 times the income of a corn farmer in Mindanao (Watkins, 1996, p.250). It is widely acknowledged that the 2002 US Farm Bill – providing US$73 billion in subsidies over the next ten years – has undermined any claims that the US supports free trade in agriculture (*Weekend Australian*, 4-5 May 2002, p.16).

Levelling the Playing Field

The fifth discursive claim of the corporate regime is belied by this undulating relation between Northern agro-exporting states and the South. While there have been concessions to states located in the Fourth World, the playing field is more like a slope, with a muscular Northern offence facing a Southern defence compromised by structural adjustment, FTAs, and WTO rules. We must add to that the power of TNCs to structure comparative advantage through the mechanisms of global sourcing.

Contesting The Corporate Regime: From Agribusiness to Agriculture

The claims of the global corporate regime are framed in 'development speak' – provision for the hungry, sustainability, and trade freedom (see McMichael, forthcoming). But the terms of engagement are new and distinct. Development is now less a purposeful national initiative, and more a reward for joining the global regime. Development is deemed to be the function no longer of states but of private capital. The WTO is charged with organizing market relations for private capital, transforming national regulatory structures and reducing the 'friction' of public economy.

In this context, development has become truly fetishized, as an independent product of the market that governs the action of states and citizens. The development paradigm always presumed a single, universal standard with which to assess well-being. Vandana Shiva observes: 'The paradox and crisis of development arises from the mistaken...identification of the growth of commodity production with providing better human sustenance for all' (1991, p.215). Aside from the cultural bias embedded in the development paradigm, its resolution to the problem of global inequality dictates more of the same, that is, more development. In a world in which environmental limits are becoming readily apparent, the problem of global inequality is as much the problem of unrestrained affluence as it is the problem of grinding poverty. In a world in which 15 per cent of the global population produces and consumes 80 per cent of the world's income, accelerating development is arguably a recipe for social and ecological disaster.

By its very nature, the crisis of development is also expressed in the proliferation of counter-movements. Some of these are direct challenges to the corporate regime – such as consumer movements concerned with labelling, food safety and fair trade; or farm worker movements concerned with pesticide use and worker security; or farmer movements concerned with protecting agriculture from agribusiness. And others are indirect challenges insofar as they mushroom in the interstices of the global economy as the basis for alternative food cultures – such as community supported agriculture, local foodsheds, and the organics movement.

Counter-movements are not simply coincidental alternatives to the corporate regime. They constitute it because they express the material and discursive conditions which the corporate agents actively seek to appropriate. For example, the global managers and the biotech corporations impose a singular and abstracted discursive and material logic on a culturally, ecologically and politically diverse world. Thus, seed patenting reduces biodiversity to monoculture under the guise of addressing the world's food needs. And, the concept of comparative advantage masquerades as an efficient allocation of global resources and benefits based on ecological and cultural endowments, but is in reality a corporate, rather than a geographical, property (Lappé, 1998, p.113). In other words, the discourse and practices of the corporate regime are not simply the assertion and management of globalization as the path to prosperity, they are also *denials* of cultural diversity, citizen's rights and biodiversity as alternative forms of sustainable practice.

Another dimension of the globalization movement that generates counter-movements is the marginalizing thrust of the global market. Where the initial, mid-Twentieth Century project of development aimed to replicate the Western model, country by country, the Twenty-first Century project of globalization is premised on specialization in the global marketplace. Necessarily regions, communities, and producers experience the uncertainty brought by the mobility of the transnational corporation and its continual reconstitution of comparative advantage. Some dependency theorists used to say that what was worse than the exploitation of foreign investment was no exploitation at all. I believe we can see beyond this aphorism. Indeed there is cautious affirmation of the blessing in disguise visited on some populations who find themselves on the margins. This movement is known as the culture of the new commons, and Mexican intellectual Gustavo Esteva (1992, p.21) observes:

> Peasants and grassroots groups in the cities are now sharing with people forced to leave the economic centre the ten thousand tricks they have learned to limit the economy, to mock the economic creed, or to refunctionalize and reformulate modern technology. The 'crisis' of the 1980s removed from the payroll people already educated in dependency on incomes and the market, people lacking the social setting enabling them to survive by themselves. Now the margins are coping with the difficult task of relocating these people. The process poses great challenges and tensions for everyone, but it also offers a creative opportunity for regeneration.

Without romanticizing this phenomenon – because it generates all manner of hideous exploitation at the same time as it enables alternative, co-operative practices – it is important to observe that the global corporate regime is highly selective and exclusive. As such, it is characterized by an *implosive* dynamic in which the triadic markets of the North American, East Asian and Northern European regions concentrate formal global economic activity. The Chairman of Citicorp distinguishes between bankable and unbankable parts of the world (Hoogvelt, 1997, p.83). This phenomenon, aided by financial globalization, and mandated by structural adjustment programs, only intensifies the Southern elites' channelling of their national wealth to the financial markets and institutions of the North. Manuel Castells points out that 'within the framework of the new informational economy, a significant part of the world population is shifting from a structural position of exploitation to a structural position of irrelevance' (cited in Hoogvelt, 1997, p.89). Irrelevance could be a virtue.

The selectivity of globalization represents both an opportunity and a danger. The danger lies in the movement towards privileging corporate over citizen rights, and the abrogation of responsibility for broad social and ecological sustainability on the part of the governments and institutions of the global system. For example, Southern critics have charged that the Global Environmental Facility, is more concerned with preserving the sink function of the world's forests and wetlands to sustain global economic activity that supports Northern lifestyles (Hildyard, 1993, pp.32-34).

Corporate greenwashing plays an important role here. It seeks to deny a problem by appropriating oppositional discourse, at the same time as it sustains the credibility of the problem of environmental and food contamination. In addition to legal intimidation through the courts, and a challenge via the WTO, the biotech industry bombards Europeans with advertising designed to undercut opposition to GM foods. However, an official EU poll demonstrates overwhelming support for labelling of GM foods, with the following favourable percentages: Austria 73 per cent, Belgium 74 per cent, Denmark 85 per cent, France 78 per cent, Germany 72 per cent, Greece 81 per cent, Ireland 61 per cent, Italy 67 per cent, Netherlands 79 per cent, Spain 69 per cent, Sweden 81 per cent, and the UK 82 per cent. While the biotech industry represents the issue as non-controversial in America, in a survey conducted by Novartis, 93 per cent of respondents favoured labelling of GM food (Goldsmith, 1998, p.314).

Just as corporate agriculture generates its own environmental opposition, so the global food system generates food scares, like mad cow, and foot and mouth disease. Food safety is increasingly threatened by global sourcing – from regions with polluted water, resource-poor safety systems, and exotic microbes unknown in the consumer countries. The Centre for Disease Control in Atlanta reports growing outbreaks of disease, some fatal, linked to imported food in the 1990s – raspberries from Guatemala, carrots from Peru, strawberries, scallions and cantaloupes from Mexico, coconut milk from Thailand, canned mushrooms from China, and Israeli snackfood. As former US President Clinton's first trade representative, Mickey Kantor said: there is 'a tension between the two goals of safety and trade. You want to open markets but not lower standards. And that's easy to say, but very, very difficult to carry out' *(New York Times,* 29 September 1997: A1, 10). No kidding!

Organic farming, as part of the counter-movement, is a case in point where there are fine lines to be drawn between corporate and alternative agricultures. Certainly sales of organic foods are burgeoning – for example it was a $5 billion industry in Canada and the US by 1997, and growing 20 per cent annually. The most recent research indicates a worldwide growth rate of some 30 per cent per annum (Cosik, 2002, p.26). A Novartis survey showed that 54 per cent of American consumers preferred organic food production as the dominant form of food production. More than 2 million American families are organic consumers, with an additional 14 million consuming natural foods (Kimbrell, 1998, p.296; Lilliston and Cummins, 1998, p.196). In 1997 the USDA attempted to redefine organic food standards to allow the use of genetic engineering, nuclear irradiation and the application of toxic sewage sludge in organic agriculture. Intensive animal farming practices, including continued reliance on antibiotics and cramped conditions for animals, would also have been deemed acceptable. Opposition was swift and unprecedented, with 27 of 40 non-governmental and state organic certifiers insisting on a uniformly high organic agricultural standard, and with 220,000 responses, 99 per cent of which denounced the proposed changes, leading to the formation of a new national organization: the Organic Consumers Association. This Association is building a movement through natural food co-ops, retail stores, farmers markets, and other community organizations, in order to contest the USDA's move to monopolize the definition of 'organic' (Lilliston and Cummins, 1998, p.200).

There is no question that the biotech industry is meeting with concerted resistance around the world (McMichael, 1998). In 1998 French farmers attacked a storage facility owned by Novartis and destroyed 30 tons of transgenic corn seed, when the French government allowed planting of GM corn. The Confederation Paysanne – an organization of European small farmers characterized this as a 'giant step toward more and more dangerous agriculture'. The Karnataka Farmers Union in Bangalore (claiming a membership of 10 million) actively resisted Cargill's attempts to patent germplasm, and more recently have turned their attention to Monsanto, ripping up and burning GM crops, and giving life to other grassroots organizations in this struggle. On 9 August, 1998, the anniversary of Gandhi's telling the British to quit India, a Monsanto Quit India campaign was launched by a group of non-farm organizations who have been mailing Quit India postcards to Monsanto's headquarters in Illinois (Kingsnorth, 1999, pp.9-10).

Schemes for sustainable agricultures crop up across the world. The Centre for Conservation of Traditional Farming Systems in Madhya Pradesh is now cultivating unirrigated wheat varieties by traditional methods, in a bid to reverse the green revolution, to reverse the socially and environmentally unsustainable impact of high-input agriculture, and to model small farming and subsistence agriculture as the alternative to big-dam-based irrigated agriculture (Nellithanam and Samiti, 1998, pp.29-33).

Conclusion

In the various ways described above, and more, the counter-movements express the crisis of development. While they do not necessarily have the same historical, cultural and philosophical point of departure, nor goal, these movements express a certain unity in rejecting or re-framing the discursive claims and material practices of the global corporate food regime. Arguably, the counter-movements are unified around reversing the marginalization of rural culture and the extreme commodification of a life force such as food.

The central issue is surely that the corporate logic is culturally reductive and unsustainable, and food may be the strongest litmus test of this. Of course, cuisines have evolved over time and across space, but we stand on a threshold beyond which the proverbial 'frankenfood' beckons to populations in what we may now call the Fast World (as opposed to the Three Worlds of development). In the North, as the corporate PR machine bulks up, the activist network will strengthen. In my opinion the latter has one key card up its sleeve: more and more consumers/parents understand the shortcomings of industrial food: 'routinely contained in nearly every bite or swallow of non-organic industrial food are antibiotics and other animal drug residues, pathogens, faeces, chemicals, toxic sludge, rendered animal protein, genetically modified organisms, chemical additives, irradiation-derived radiolytic chemical by-products, and a host of other hazardous allergens and toxins' (Lilliston and Cummins, 1998, p.196).

The crisis of development is not simply the crisis of a model, but a philosophical and ethical crisis with a very hard edge of growing inequality. The upward redistribution of wealth is accompanied by the arrogance of power of the wealthy as they reconfigure the world in the service of short-run profit. It seems to me that over the long haul citizens, workers and consumers will not subscribe to the market fetish, as some higher authority than the polity, as they experience social polarization and the erosion of social institutions and stable habitats and communities. The crisis of development is a crisis of institutions of governance also. The globalization project seeks to strengthen states and multilateral institutions as corporate entities but weaken boundaries – both national and, in this case, biological. The counter-movement, on the other hand, is exploring alternative political and social forms – from global citizen networks to community-level organization to what Wolfgang Sachs terms 'cosmopolitan localism', that is local activism situated within its world-historical context.

Much is made of the alternative scale of action of contemporary counter-movements. However, given the scenario outlined here, I believe that such movements cannot avoid engaging with policy-making institutions, and addressing the transformation of states into corporate entities – this process is progressively shrinking democratic political space. One obvious example of such engagement is questioning the legality and ethics of bio-engineering. Another is the discursive dispute over the definition of 'organic', which has serious material consequences. And the terrain of this dispute in the US directly implicates the global regime insofar as how the US government defines organic affects its trade relations. A confidential USDA memo reported in *Mother Jones* magazine in April 1998 remarked: 'Few if any existing [organic] standards permit GMOs [genetically modified organisms], and their inclusion could affect the export of US Grown organic product. However, the Animal and Plant Health Inspection Service and the Foreign Agricultural Service are concerned that our trading partners will point to a USDA organic standard that excludes GMOs as evidence of the department's concern about the safety of bio-engineered commodities' (cited in Lilliston and Cummins, 1998, p.197).

Finally, for the majority of the world's population, food is not just an item of consumption, it is actually a way of life. It has deep material and symbolic power. And because it embodies the links between nature, human survival and health, culture and livelihood, it will, and has already, become a focus of contention and resistance to a corporate takeover of life itself.

References

Barndt, D. (1997), 'Bio/cultural Diversity and Equity in Post-NAFTA Mexico (or: Tomasita Comes North While Big Mac Goes South)' in J. Drydyk, and P. Penz (eds), *Global Justice, Global Democracy*, Fernwood Publishing, Winnipeg/Halifax, pp. 55-59.

Bruno, K. (1998), 'Monsanto's Failing PR Strategy', *The Ecologist*, vol. 28(5), pp. 287-93.

Cosik, M. (2002), 'Eat, Drink and Be Wary', *Weekend Australian Magazine*, 12-13 January, p. 26.

Esteva, G. (1992), 'Development', in W. Sachs (ed.), *The Development Dictionary*, Zed Books, London.

Gill, S. (1992), 'Economic Globalization and the Internationalization of Authority: Limits and Contradictions', *Geoforum*, vol. 23(3), pp. 269-83.

Goldsmith, Z. (1998), 'Who Are the *Real* Terrorists?', *The Ecologist,* vol. 28(5), pp. 312-17.

Gorelick, S. (2000), 'Facing the Farm Crisis', *The Ecologist*, vol. 30(4), pp. 28-32.

Greenfield, G. (1999), *The WTO, the World Food System, and the Politics of Harmonised Destruction,* www.labournet.org/discuss/global/wto.html.

Heffernan, B. (1999), *Consolidation in the Food and Agriculture System*, Report to the National Farmers Union, US.

Hildyard, N. (1993), 'Foxes in Charge of Chickens', in W. Sachs (ed.), *Global Ecology*, Zed Books, London.

Hindmarsh, R. and Lawrence, G. (eds)(2001), *Altered Genes II: The Future?* Scribe, Melbourne.

Hoogvelt, A. (1997), *Globalization and the Postcolonial World: The New Political Economy of Development*, Macmillan, London.

Kimbrell, A. (1998), 'Why Biotechnology and High-Tech Agriculture Cannot Feed the World', *The Ecologist*, 28(5), pp. 294-298.

Kingsnorth, P. (1999), 'India Cheers While Monsanto Burns', *The Ecologist*, vol. 29(1), pp. 9-11.

Kneen, B. (1995), *Invisible Giant: Cargill and its Transnational Strategies*, Pluto, London.

Lappé, F., Collins, J., Rosset, P. with Esparza, L. (1998), *World Hunger: Twelve Myths*, second edition, Grove Press, New York.

Lappé, M., and Bailey, B. (1998), *Against the Grain. Biotechnology and the Corporate Takeover of Your Food*, Common Courage Press, Monroe, ME.

LeQuesne, C. (1997), 'The World Trade Organisation and Food Security', *Talk to UK Food Group*, London, July 15.

Lilliston, B. and Cummins, R. (1998), 'Organic Vs 'Organic': The Corruption of a Label', *The Ecologist*, 28(4), pp.195-200.

McCalla, A. (1999), 'World Agricultural Directions: What Do They Mean For Food Security?' *Paper presented at Cornell University, Sponsored by Cornell International Institute for Food, Agriculture and Development*, March 30, 1999.

McMichael, P. (1995), 'The "New Colonialism": Global Regulation and the Restructuring of The Inter-State System', in D. Smith and J. Borocz (eds) *A New World Order? Global Transformations in the Late Twentieth Century*, Greenwood Press, Westport, pp. 37-56.

McMichael, P. (1996; 2000), *Development and Social Change: A Global Perspective*, Thousand Oaks, Pine Forge Press.

McMichael, P. (1997), 'Rethinking Globalisation: The Agrarian Question Revisited', *Review of International Political Economy*, vol. 4(4), pp. 630-62.

McMichael, P. (1998), 'Global Food Politics', in F. Magdoff, J. Foster and F. Buttel (eds) in 'Hungry for Profit', *Monthly Review*, vol. 50(3), pp. 97-111.

McMichael, P. (forthcoming) 'Biotechnology and Food Security: A Timely Connection?' in L. Beneria and S. Bisnath (eds) *Global Tensions: Challenges and Opportunities in the World Economy*, Routledge, New York.

McMichael, P. and Lawrence, G. (2001), 'Globalising Agriculture: Structures of Constraint for Australian Farming', in S. Lockie, and L. Bourke, (eds) *Rurality Bites: The Social and Environmental Transformation of Rural Australia*, Pluto, Sydney, pp. 153-164.

Mendelson, J. (1998), 'Roundup: The World's Biggest-Selling Herbicide', *The Ecologist*, vol. 28(5), pp. 270-75.

Nellithanam, R., Nellithanam, J., and Samiti, S. (1998), 'Return of the Native Seeds', *The Ecologist*, vol. 28(1), pp. 29-33.

New York Times, 29 September 1997, A1, 10.

Rifkin, J. (1998), *The Biotech Century: Harnessing the Gene and Remaking the World*, Tarcher/Putnam, New York.

Rist, G. (1997), *The History of Development. From Western Origins to Global Faith*, Zed, London and New York.

Shiva, V. (1991), *The Violence of the Green Revolution*, Zed Books, London.

Watkins, K. (1996), 'Free Trade and Farm Fallacies. From the Uruguay Round to the World Food Summit', *The Ecologist*, vol. 26(6), pp. 244-255.

Weekend Australian, 4-5 May 2002, p.16.

PART II
THE RESTRUCTURING OF LOCAL AGRICULTURE

Chapter 6

Assessing the Neoliberal Experiment in Antipodean Agriculture

Hugh Campbell and Geoffrey Lawrence

Introduction: Neoliberalism Through 'Antipodean Eyes'

Events at Seattle in November 1999 and in Genoa in 2001 underscore one of the enduring conflicts within the global adoption (or rejection) of neoliberal trade and economic policies. The reality of two decades of General Agreement on Tariffs and Trade (GATT) negotiations, the completion of the Uruguay Round, and the turmoil that accompanied the World Trade Organization's (WTO's) 'millennial round' in Seattle, is that neoliberal trade and economic policies have been applied inconsistently across a limited number of economic sectors. Another reality is that the so-called 'dry' policies of economic rationalism are being actively resisted by peoples in various parts of the global economy. Unevenness and conflict have often been at their most acute in the political regulation of agriculture.

It is in this context that the political 'purity' of the Cairns group of free-trading agricultural exporters has been so interesting. No nations have been more vigorous advocates of subjecting agriculture to neoliberal reform than Australia and New Zealand. Consequently, the experience of antipodean agriculture has become something of a *cause celebre* or test case of agriculture under neoliberalism. From both outside (see Cloke, 1989, 1996; Drummond *et al.*, 2000) and within (see Lawrence, 1987; 1996a; 1996b; Campbell, 1994; Burch *et al.*, 1999; Gray and Lawrence, 2001) these countries, rural sociologists have attempted to understand the changes occurring in agriculture and rural society.

In this chapter we outline the way in which the notion of crisis developed by second wave French Regulation Theory scholars like Moulaert and Swyngedow (1989), and Peck and Tickell (1992; 1994a; 1994b; 1995; Tickell and Peck, 1992) can be used to understand the nature of recent changes in antipodean agriculture and the role of neoliberalist policies in that crisis (see also Campbell and Coombes, 1999). Moulaert and Swyngedow (1989) outline a regulationist understanding of crisis by suggesting that the movement from stable regime to crisis is characterized by a number of phases:

- Short 'conjunctural' crises requiring minor adjustments (for instance, incremental technological changes, expanding spatial divisions of labour, and institutional adjustments);
- Structural crises (or crises of a particular mode of development) leading to qualitative changes in the organization of the accumulation process; and
- Crises resulting from fundamental contradictions in the capitalist mode of production itself (Moulaert and Swyngedow, 1989, p.329).

The following account will trace the history of neoliberalism in the Australian and New Zealand cases. Our analysis will outline the way in which a stable post-war agrarian (and national) order was increasingly disrupted and fractured during the 1970s. As the *conjunctural* crisis intensified, the neoliberal experiment in the Antipodes emerged as a potential solution – only to plunge agriculture into a full *structural* crisis. The results were ongoing instability and the inability of neoliberalism to 'fix' or stabilize capital accumulation in a new national or global regime.

Australia and New Zealand: From Stability to Crisis

European style agriculture, imported to Australia and New Zealand following the conquests of Indigenous inhabitants from the late 18th to the late 19th Centuries, conformed to a 'settler capitalist' mode (McMichael, 1984). Produce left Australian and New Zealand shores as largely unprocessed primary products such as wool, lamb, beef and wheat. These formed important components of the imperial food order – the First Food Regime (Friedmann and McMichael, 1989). From the turn of the century, British investment – alongside protectionist policies and entrenched trade unionism in the colonies – ensured that in both Australia and New Zealand, the benefits of economic growth generally reached all classes. Rural producers were advantaged by the dependency relationship that ensured Britain's industrializing needs would be met. This pattern was retained through two major depressions (1890s and 1930s) and experienced its boom period in both Australia and New Zealand in the period from 1945 to 1973. Throughout the boom, the stability of this economic configuration was maintained through a mantle of protection, subsidization, price support, research and state regulation.

This long boom of prosperity and stability after the Second World War began to unravel in the late 1960s and enter what Moulaert and Swyngedow (1989) term a 'conjunctural crisis'. Initially indicated, in 1967, by a major downturn in the price of wool, the 'solutions' that state intervention provided to ensure ongoing accumulation and stability became increasingly unviable. In 1973, Britain entered the Common Market, effectively ending its colonial food ties with Australia and New Zealand, and the oil shock and the collapse of the Bretton Woods agreement impacted negatively on trade. Both Australian and New Zealand governments attempted a range of interventions to assist agriculture. These attempts to find interventionist solutions to a conjunctural crisis eventually ended in the mid-1980s as the governments of both nations proceeded with a radical experiment to find solutions via neoliberal policy.

While many policy advocates of the time suggested that neoliberalism would provide a coherent new regulatory regime, we contend that the neoliberal experiment simply moved the agricultural sectors of Australia and New Zealand from the last stages of a conjunctural crisis into a full blown structural crisis. This provoked a range of further 'crisis experiments' in an attempt to secure stable accumulation under a neoliberal policy framework (Campbell and Coombes, 1999). Cloke (1989) and Cloke and Le Heron (1994) argue that the reasons for the ease with which the Antipodean nations were captured by neoliberal ideology were the small size of the political elite – and their desire to conform to the demands of their major trading partners – and the alarming nature of the emerging economic problems in each country. In both countries, the relatively low levels of state support for agriculture – relative, that is, to Britain and Europe – throughout the long boom meant that farmers could not cling tenaciously to subsidization and state support to the degree evidenced elsewhere. In fact they are viewed as having willingly cooperated in their own demise – accepting policies that would 'restructure' many of their fellow producers out of existence (Lawrence, 1987; Campbell, 1994).

Various policy initiatives were introduced to expose agriculture more directly to global conditions – as part of the wider neoliberal experiment to restructure the economies of both countries. Since the early 1980s governments in Australia and New Zealand have removed the many interventions characteristic of the stable regime. They reduced or entirely removed tariffs; floated their currencies and lifted foreign exchange controls; deregulated finance and banking, and engaged in monetarist strategies to control inflation. In agriculture, they abandoned or reduced their commitment to statutory marketing; removed most of the direct subsidies to farming; eliminated or severely curbed price supports; dismantled the favourable tax regime for farmers; reduced government sponsored research activity; and introduced user-pays principles for extension and information services (Lawrence, 1987; Fairweather, 1992; Tyler and Lattimore, 1990; Britton *et al.*, 1992; Le Heron, 1993).

While we see the historical events that preceded the adoption of neoliberalism as clearly being explicable in the terms used by French Regulation Theory, this is not a majority view. A host of neoliberalists in Australia and New Zealand have seen the mid-80s as Year Zero in a bold attempt to reinvigorate antipodean agriculture. In the ensuing 15 years, most commentators have found it difficult to argue that the reform process has delivered the promised solutions. Having assumed that the only way agriculture in the Antipodes would prosper would be through a more globalized free trade regime (ABARE, 1997a), neoliberal economists have sought to excuse the lack of economic growth on the specific nature of traded agricultural commodities, low commodity prices, unfair world competition, the relative immovability of assets 'fixed' in farming, or the need for even more liberalization (see Sandrey and Reynolds, 1990; Malcolm *et al.*, 1996; Godden, 1997). What the neoliberals have concluded is that with the:

> economic forces that farmers confront...there will eventually either have to be fewer farmers...or poorer farmers. The third option is for farmers to be subsidized by taxpayers and consumers. This option has been widely adopted in the USA and Western Europe (Malcolm *et al.*, 1996, p.58).

Since the early 1980s, only the first two scenarios have been considered politically acceptable in Australia and New Zealand: the third is viewed as undesirable because it involves state intervention, argued to be both a market distortion and an unfair imposition upon the wider community (Gow, 1994; Godden, 1997). This view maintains a position of hegemony in political discourse in Australia and New Zealand, and appears to take no cognisance of increasing evidence that Antipodean agriculture remains in crisis. The intensification of agriculture, as well as persistent low incomes in farming, have been implicated in continued environmental degradation – particularly in Australia (Conacher and Conacher, 1995; Vanclay and Lawrence, 1995). Stress levels and general social malaise continue to be features of contemporary farming (Fairweather, 1989; Gray *et al.*, 1993). Yet, there are quite obvious limits to the ability of the smaller nations to persuade other, competing nations – such as Britain and US – to remove subsidization from farming so as to create a 'level playing field'.

Structural Crisis in Australia and New Zealand Agriculture

An alternative explanation to that provided by neoliberal interpretations of the failure of the 1980s' reforms is that the mid-1980s reflected the movement from a 'conjunctural crisis' to a more pernicious 'structural crisis' (Moulaert and Swyngedow, 1989). It argues that neoliberalism was not 'the only alternative' as advocated by the New Zealand and Australian Treasuries, and farm organizations (Liepins and Bradshaw, 1999), but an 'experiment' to solve an emerging crisis that appears to have actually exacerbated rather than ameliorated that crisis. The reasons for claiming a *structural* crisis in agriculture will be assessed in four areas that we consider to be consistent with the understandings of second wave French Regulation Theory. These are:

- Destabilization of the economic basis of agricultural production;
- Significant changes in the social structure of farm production;
- Disruption of the prevailing ideologies which constituted normal and legitimate farming practices; and
- An overall tendency towards instability and unsustainable activities.

These are described below:

Destabilization of the Economic Basis of Agriculture

The immediate effect of economic liberalization was a decline in farm revenues (Johnson, 1991). Removal of price supports for sheep (and immediate deregulation of some domestic producer boards) in New Zealand caused a plunge in farm revenues. The market collapse in 1986 resulted in some sheep being sold for only NZ$2 per head (Campbell, 1994). As in Australia, this cost did not even cover the transport of these animals to sale. The removal of producer-board intervention prices exposed farmers to world prices at a time of falling global demand. The suddenness of deregulation made this particularly painful in New Zealand – but similar effects were also emerging in Australia with collapses in wool, wheat and beef prices (Lawrence, 1987).

Falling farm revenues were accompanied by severe declines in land value as prices readjusted to both the reduced earning power of staple farm products and the removal, in 1985, of subsidized loans to purchase farmlands (Johnston and Sandrey, 1990). Demand for farmland also fell due to rigorous monetary policy in both Australia and New Zealand which forced up interest rates in an attempt to control inflation. By 1987, farmland prices had declined so far that for many farm types equity levels fell below 40 per cent, while for some farm types negative equity occurred, leading to severe conflict with lending institutions (Campbell, 1994). The fundamental economic features of the structural crisis were, therefore, a diminished ability to generate revenue from farming, collapsing equity, and a destabilization of the farm/finance relationship.

Changes to the Social Structure of Agricultural Production

Throughout this crisis there were significant changes to the social structure of farm production. In New Zealand, personal expenditures for farm families dropped dramatically (Campbell, 1994) and they have never fully recovered. Farm labour patterns also changed, with increased levels of family labour on farms and an increase in the level of off-farm work (Benediktsson *et al.*, 1990; Britton *et al.*, 1992; Le Heron *et al.*, 1994), although the trend towards off farm work had already begun and was merely intensified through the crisis (see Coombes and Campbell, 1996). Pluriactivity also intensified, yet the ability of farms to diversify into new economic activities was linked to wider, 'post-modern' options in the tourism and leisure industries – something particularly evident in coastal Australia (Lawrence, 1995). Throughout the major part of the crisis period, self-exploitation by family members increased markedly (Gray *et al.*, 1993; Campbell, 1994) and continues to be the key to family-farm economic viability today (see Gray and Lawrence, 2001).

Farm sizes increased throughout most areas, while farm divisions occurred in particular pockets – those where producers could take advantage of city people's desires for an alternative lifestyle by subdividing their properties. In New Zealand, a disappearing middle in farm sizes began to occur after 1984 with the main changes being an increase in the average size of family farms and an increase in subdivision and establishment of lifestyle properties in peri-urban areas (Fairweather, 1992). In both Australia and New Zealand there was an exacerbation of the trend toward long-term population loss in many rural areas – including the growth in the size of properties and the increasing linkages of larger properties with transnational capital (Burch *et al.*, 1996; ABARE, 1997b; Burch *et al.*, 1999). While under the previous regime, the state had intervened to underwrite the social and economic infrastructure of rural communities, the move towards neoliberalism saw an abandonment of any 'rural policy' by both New Zealand and Australia (Stehlik and Lawrence, 1996; Gray and Lawrence, 2001).

Undermining Established Institutions, Norms and Practices

Within a regulationist interpretation, an important aspect of a structural crisis is the way that established norms, practices, and ideologies underpinning farming are destabilized or delegitimized (Coombes and Campbell, 1996; Campbell, 1999). This aspect of crisis has been neglected in prior accounts. The farm debt crisis in New Zealand agriculture was the direct consequence of a regime of state supported production, high debt levels and guaranteed prices prior to 1984, suddenly moving to deregulation and uncertainty by 1986/87. Campbell (1994; 1999) indicates that central ideological concepts in the farm/finance relationship like 'equity', 'credit worthiness' and who constituted a 'good farmer' were completely undermined by the collapse in land prices and new ideologies of business practice emerging in an internationalized finance sector. Farmers could not understand their circumstances in terms of the ideological interpretations associated with the old regime. Throughout this crisis, different groups including banks, lawyers, accountants, farm families, local farmer groups and stock firms contested the fate of a myriad of farms – a process which remains unstable in both Australia and New Zealand, as currency exchange rates, interest rates, and credit ratings continue to fluctuate – and a new set of ideological meanings around farm financing and practice have failed to emerge.

This is not an isolated example of how prevailing ideologies become disrupted in a structural crisis. In other areas of farm activity, old ideologies of the gender division of labour (Taylor and McCrostie Little, 1995), the role of paid farm labour (Coombes and Campbell, 1996), notions of farm succession (Keating and Little, 1991; 1994), relations between farms and stock firms (Campbell, 1994), and other ideological givens of prior farm activity were completely disrupted during the crisis period. This might be described as a process of detraditionalization of rural society leading to uncertainties about change and direction, and to the demise of numerous social institutions and the growth of political instability (Gray, 1996; Heelas, 1996; Luke, 1996; Gray and Lawrence, 2001). Political instability is clearly exemplified in recent events in rural Australia with the scapegoating of racial and ethnic minorities who are blamed for the economic problems of the nation (Gray, 1996; Cheers 1998; Lawrence *et al.*, 1999).

Tendency Towards Instability and Unsustainable Activities

All the above effects would not be so serious if they were merely a 'time of pain' in the transition to a new stable order of farm prosperity, as the neoliberal rhetoric suggests. The reality is that farming is now more unstable and insecure than it was under the old regime – even as it was entering its crisis post-1973. During the neoliberal experiment many farm producers have tried to diversify in circumstances where exporting was occurring in an international environment of unstable commodity prices and rapidly changing niche opportunities and also where revenues from export activities suffered continuing destabilization as a result of currency fluctuations. In New Zealand, many pastoral producers experimented with beef production – only to see the beef price fall dramatically. Without any endogenous regulation of the export sector, producers are directly affected by international price instability. The exception proves this rule. The most sustained growth in New Zealand agriculture has occurred within dairying – an industry which has held out against neoliberalism through cooperative production and militant farmer protection of statutory control over marketing activities thereby maintaining a stable industry structure throughout the crisis.

Finally, both R & D and environmental policies in both countries have failed to engage with environmental sustainability or are completely contradictory – presaging ongoing instability in the agriculture and environment relationship. The reduction of the activities of the state as a provider of research and innovation in agriculture, through the substantial downsizing of the Ministry of Agriculture and Fisheries in New Zealand and the State Departments of Agriculture in Australia, has resulted in corporate agri-business having an increased role in innovating and developing new products and production systems (Burch *et al.*, 1996). The ephemeral commitment of corporations to such developments, only insofar as they are profitable, leads to a degree of instability (Campbell and Coombes, 1999). This becomes most clear in regard to agriculture's impact on the environment. Here, New Zealand and Australia have adopted different strategies in response to environmental concerns.

In New Zealand, despite the rhetoric of the Resources Management Act (1991), mainstream agricultural production is exempt from most sanctions on unsustainable resource management. Instead, the government has directed its energies away from issues of the broader public good in agricultural/environmental sustainability and has only supported the development of 'sustainable' systems insofar as they can produce lucrative export products. Despite the modest environmental gains inherent in such a strategy, the limitations expose 'sustainable development' to the instabilities and vagaries of world product markets, while falling well short of a thoroughgoing governmental commitment to environmental sustainability in agriculture (Campbell and Coombes, 1999). Even the marginalized sustainable agriculture movement – which may have potentially formed one base of resistance – exists in a relationship of 'dependent reproduction' with mainstream, liberalized agricultural exporting (Coombes and Campbell, 1998).

In contrast, the state in Australia has attempted to address the consequences of past rounds of environmental degradation within a context of neoliberal agricultural production – resulting in a range of initiatives aimed at producer (not state) responsibility for the environment. It funds a number of important voluntary initiatives such as Landcare, where farming and community-based groups receive funding to undertake practical measures to restore degraded environments (Lockie and Vanclay, 1997) and to manage catchments in an integrated manner (Natural Heritage Trust, 1997; AFFA, 1999). The question of whether Landcare and catchment management are little more than ideologies that obscure the continuation of entrenched and unsustainable production practices has been raised by a number of authors (Vanclay and Lawrence, 1995; Bailey, 1997; Lockie, 1997). The suspicion is that continued environmental degradation will occur as financially stretched producers continue to 'mine' the soil, and overstock and overgraze, in the hope of improved prices in future years (Vanclay and Lawrence, 1995; Hamilton, 1996; Gray and Lawrence, 2001). Consequently, the overall commitment by the state to a crisis experiment of neoliberalism has resulted in truncated or tokenistic attempts to meet environmental concerns in agriculture. While the strategies differ between Australia and New Zealand the contradictions are clearly evident in both.

Conclusion

The argument of this chapter is that the agricultural sectors of both Australia and New Zealand moved, during the mid 1980s, from a conjunctural crisis to a structural crisis. The result has been a period of prolonged instability as neoliberalism has failed to deliver stable growth, to improve incomes or to improve the prospects for environmental sustainability. Despite some suggestions that a successor regime might be emerging we would contend that crisis and instability are still the main characteristics of Antipodean agriculture. The state has curtailed its intervention in agricultural production in both countries, and attempts to form an environmental policy for agriculture have been truncated and subordinated to the overall drive towards neoliberalism, while any vestiges of a rural policy by both countries has been largely abandoned. Importantly, localized resistance and/or amelioration of the effects of the neoliberal experiment have been slow to materialize.

Note

The Australian and New Zealand 'case study' in this paper has been extracted and adapted from Drummond, I., Campbell, H., Lawrence, G. and Symes, D. (2000), 'Contingent or Structural Crisis in British Agriculture', *Sociologia Ruralis*, vol. 40(1), pp.111-127. We thank Dr Henk de Haan, editor of *Sociologia Ruralis*, and Ian Drummond and David Symes for permission to use this material.

References

Agriculture, Fisheries and Forestry – Australia (AFFA)(1999), *Managing Natural Resources in Rural Australia for a Sustainable Future: A Discussion Paper for Developing a National Policy*, AFFA, Canberra.

Australian Bureau of Agriculture and Resource Economics (ABARE)(1997a), 'Advantages of Trade Reform', *ABARE Current Issues*, July, 5, pp. 1-8.

Australian Bureau of Agriculture and Resource Economics (ABARE)(1997b), 'Changing Structure of Farming', *ABARE Current Issues*, June, 4, pp. 1-8.

Bailey, M. (1997), *Landcare: Myth or Reality?* in S. Lockie and F. Vanclay, (eds), *Critical Landcare*, Key Papers Series 5, Centre for Rural Social Research, Charles Sturt University, pp. 129-142.

Benediktsson, K., Manning, S., Moran, W. and Anderson, G. (1990), *Participation of Raglan County Farm Households in The Labour Force*, Occasional Paper no. 27, Department of Geography, University of Auckland.

Britton, S., Le Heron, R. and Pawson, E. (1992), *Changing Places in New Zealand: A Geography of Restructuring*, New Zealand Geographical Society, Christchurch.

Burch, D., Rickson, R. and Lawrence, G. (eds), (1996), *Globalization and Agri-food Restructuring: Perspectives from the Australasia Region*, Avebury, Aldershot.

Burch, D., Lawrence, G. and Goss, J. (eds), (1999), *Restructuring Global and Regional Agricultures: Transformation in Australasian Economies and Spaces*, Ashgate, Aldershot.

Campbell, H. (1994), *Regulation and Crisis in New Zealand Agriculture: The Case of Ashburton County, 1984-1992*, Unpublished Doctoral Thesis, Charles Sturt University, Australia.

Campbell, H. (1996), 'The Restructuring of Organic Agriculture in New Zealand: "Corporate Greening", Transnational Corporations and Sustainable Agriculture', in D. Burch, G. Lawrence and R. Rickson (eds), *Globalisation and Agri-food Restructuring: Perspectives from Australia and New Zealand*, Avebury, Aldershot, pp. 153-169.

Campbell, H. (1999), 'Crisis in the Farm/finance Relationship: a Regulationist Analysis of Evidence from Ashburton County', in *Proceedings of the 18th Geographical Society Conference*, Christchurch.

Campbell, H. and Coombes, B. (1999), '"Green Protectionism" and Organic Food Exporting from New Zealand: Crisis Experiments in the Breakdown of Fordist Trade and Agricultural Policies', *Rural Sociology, vol.* 64(2), pp. 302-319.

Cheers, B. (1998), *Welfare Bushed: Social Care in Rural Australia,* Ashgate, Aldershot.

Cloke, P. (1989), 'State Deregulation and New Zealand's Agricultural Sector', *Sociologia Ruralis*, vol. 24(1), pp. 34-48.

Cloke, P. (1996), 'Looking Through European Eyes? A Re-Evaluation of Agricultural Deregulation in New Zealand', *Sociologia Ruralis*, vol. 36(3), pp. 307-330.

Cloke, P. and Le Heron, R. (1994), *Agricultural Deregulation: The Case of New Zealand*, in P. Lowe, T. Marsden and S. Whatmore, (eds), *Regulating Agriculture*, David Fulton, London, pp. 104-126.

Conacher, A. and Conacher, J. (1995), *Rural Land Degradation in Australia*, Oxford, Melbourne.

Coombes, B. and Campbell. H. (1996), 'Pluriactivity in (and Beyond?) a Regulationist Crisis', *New Zealand Geographer*, vol. 52(2), pp. 11-17.

Coombes, B. and Campbell. H. (1998), 'Dependent Reproduction of Alternative Modes of Agriculture: Organic Farming in New Zealand', *Sociologia Ruralis*, vol. 38(2), pp. 127-145.

Drummond, I., Campbell, H., Lawrence, G. and Symes, D. (2000), 'Contingent or Structural Crisis in British Agriculture', *Sociologia Ruralis*, vol. 40(1), pp. 111-127.

Fairweather, J. (1989), *Some Recent Changes in New Zealand Society*, AERU Discussion Paper No. 124, Agribusiness and Economics Research Unit, Lincoln University, New Zealand.

Fairweather, J. (1992), *Agrarian Restructuring In New Zealand*, Research Report No. 213, Agribusiness and Economics Research Unit, Lincoln University, Canterbury.

Friedmann, H. and McMichael, P. (1989), 'Agriculture and the State System: The Rise and Decline of National Agricultures, 1870 to the Present', *Sociologia Ruralis*, vol. 29, pp. 93-117.

Godden, D. (1997), *Agricultural and Resource Policy: Principles and Practices,* Oxford University Press, Melbourne.

Gow, J. (1994), 'Farm Structural Adjustment – An Everyday Imperative', *Rural Society*, vol. 4(2), pp. 9-13.

Gray, I. (1996), 'The Detraditionalization of Farming', in D. Burch, R. Rickson and G. Lawrence, (eds), *Globalization and Agri-food Restructuring: Perspectives from the Australasia Region*, Avebury, Aldershot, pp. 91-103.

Gray, I., Lawrence, G. and Dunn, T. (1993), *Coping with Change: Australian Farmers in the 1990s*, Centre for Rural Social Research, Charles Sturt University, Wagga Wagga.

Gray, I. and Lawrence, G. (2001), *A Future for Regional Australia: Escaping Global Misfortune,* Cambridge University Press, Cambridge.

Hamilton, C. (1996), *Economic Rationalism in the Bush*, Paper Presented to the Department of Primary Industries and Energy, Canberra, 19 November.

Heelas, P. (1996), 'On Things Not Being Worse, and the Ethic of Humanity', in P. Heelas, S. Lash, and P. Morris (eds), *Detraditionalization: Critical Reflections on Authority and Identity*, Blackwell, Massachusetts, pp. 200-222.

Johnson, R. (1991), 'Current Changes in New Zealand Agriculture: A Review', *Review Of Marketing and Agricultural Economics*, vol. 59(2), pp. 130-148.

Johnston, W. and Sandrey, R. (1990), 'Land Markets and Rural Debt', in R. Sandrey and R. Reynolds, (eds) *Farming Without Subsidies: New Zealand's Recent Experience*, GP Books, Wellington.

Keating, N. and Little, H. (1991), *Generations in Farm Families: Transfer Of The Family Farm In New Zealand*, Research Report No. 208, Agribusiness and Economics Research Unit, Lincoln University, New Zealand.

Keating, N. and Little, H. (1994), 'Getting Into It: Farm Roles and Careers of New Zealand Women', *Rural Sociology*, vol. 59(4), pp. 720-736.

Lawrence, G. (1987), *Capitalism and the Countryside: the Rural Crisis in Australia*, Pluto, Sydney.

Lawrence, G. (1995), *Futures for Rural Australia: From Agricultural Productivism to Community Sustainability*, Inaugural Address, Central Queensland University, Rockhampton.

Lawrence, G. (1996a), 'Contemporary Agri-Food Restructuring: Australia and New Zealand', in D. Burch, G. Lawrence and R. Rickson (eds), *Globalisation and Agri-food Restructuring: Perspectives from Australia and New Zealand*, Avebury, Aldershot, pp. 45-72.

Lawrence, G. (1996b), 'Rural Australia: Insights and Issues from Contemporary Political Economy', in G. Lawrence, K. Lyons and S. Momtaz (eds), *Social Change in Rural Australia*, RSERC, Rockhampton, pp. 332-349.

Lawrence, G., Gray, I. and Stehlik, D (1999), 'Changing Spaces: The Effects of Macro-social Forces on Regional Australia', in C. Kasimis and A. Papadopoulos (eds), *Local Responses to Global Integration: Toward a New Era of Rural Restructuring*, Ashgate, Aldershot, pp. 63-87.

Le Heron, R. (1993), *Globalised Agriculture: Political Choice,* Pergamon Press, Oxford.

Le Heron R., Roche M. and Johnston T. (1994), 'Pluriactivity: an Exploration of Issues with Reference to New Zealand's Livestock and Fruit Agro-commodity Systems', *Geoforum,* vol. 25(2), pp. 155-171.

Liepins, R and Bradshaw, B (1999), 'Neoliberal Agricultural Discourse in New Zealand: Economy, Culture and Politics Linked', *Sociologia Ruralis,* vol. 39(4).

Lockie, S. (1997), 'What Future Landcare? New Directions Under Provisional Funding', in S. Lockie and F. Vanclay (eds), *Critical Landcare*, Centre for Rural Social Research, CSU, Wagga Wagga, pp. 227-238.

Lockie, S. and Vanclay, F. (eds)(1997), *Critical Landcare*, Centre for Rural Social Research, CSU, Wagga Wagga.

Luke, T. (1996), 'Identity, Meaning and Globalization: Detraditionalization in Postmodern Space-Time Compression', in P. Heelas, S. Lash and P. Morris (eds), *Detraditionalization: Critical Reflections on Authority and Identity*, Blackwell, Massachusetts, pp. 109-133.

Malcolm, B., Sale, P. and Egan, A. (1996), *Agriculture in Australia: an Introduction*, Oxford, Melbourne.

McMichael, P. (1984), *Settlers and the Agrarian Question: Foundations of Capitalism in Colonial Australia*, Cambridge University Press, New York.

Moulaert, F. and Swyngedouw, E. (1989), 'A Regulation Approach to the Geography of Flexible Production Systems', *Environment and Planning D: Society and Space*, vol. 7, pp. 327-345.

Natural Heritage Trust (1997), *Guide to Community Group Applications*, Natural Heritage Trust, Canberra.

Peck, J. and Tickell, A. (1992), 'Local Modes of Social Regulation? Regulation Theory, Thatcherism and Uneven Development', *Geoforum,* vol. 23(3), pp. 347-363.

Peck, J. and Tickell, A. (1994a), 'Searching for a New Institutional Fix: the After-Fordist Crisis and the Global-Local Disorder', in A. Amin (ed.), *Post-Fordism: A Reader*, Basil Blackwell, Oxford.

Peck, J. and Tickell, A. (1994b), 'Jungle Law Breaks Out: Neoliberalism and Global-Local Disorder', *Area,* vol. 26(4), pp. 317-326.

Peck, J. and Tickell, A. (1995), 'The Social Regulation of Uneven Development: "Regulatory Deficit", England's South East, and the Collapse of Thatcherism', *Environment and Planning A*, vol. 27, pp. 15-40.

Sandrey, R. and Reynolds, R. (1990), *Farming Without Subsidies*, GP Books, Wellington.

Stehlik, D. and Lawrence, G. (1996), '"Rural Policy" a Contradiction in Terms? The Example of Aged Care Policy in Australia', *Journal of the Australian Studies Institute*, vol. 3(1), April, pp. 91-108.

Taylor C. and McCrostie Little H. (1995), *Means of Survival? A Study of Off-Farm Employment*, Taylor Baines, Christchurch.

Tickell, A. and Peck, J. (1992), 'Accumulation, Regulation and the Geographies of Post-Fordism: Missing Links in Regulationist Research', *Progress in Human Geography,* vol. 16(2), pp. 190-218.

Tyler, L. and Lattimore, R. (1990), 'Assistance to Agriculture', in R. Sandrey and R. Reynolds, (eds), *Farming Without Subsidies: New Zealand's Recent Experience*, GP Books, Wellington.

Vanclay, F. and Lawrence, G. (1995), *The Environmental Imperative: Eco-social Concerns for Australian Agriculture*, CQU Press, Rockhampton.

Chapter 7

Resisting Integration in the Global Agro-Food System: Corporate Chickens and Community Controversy in Texas

Douglas H. Constance, Alessandro Bonanno, Caron Cates, Daniel L. Argo and Mirenda Harris

Introduction

One of the major features of globalization is *global sourcing* (Bonanno and Constance, 1996; McMichael, 1996). Global sourcing refers to the corporate global search for favourable socio-economic conditions that include a suitable political and business regulatory climate, docile and inexpensive labour, and available and adequate natural resources. Global sourcing has been particularly visible in the agro-food sector in the US through the emergence of confined animal feeding operations. Often promoted by transnational corporations as forms of capital penetration in economically depressed rural areas, these ventures have been particularly visible in regions where limited environmental, community wellbeing and labour regulations attract external corporate investments. While receiving support from political elites, these strategies have engendered widespread resistance. This crisis escalated to the point to delegitimize corporate agro-food agendas and existing Federal and regional regulations. It therefore required the intervention of public agencies such as the Department of Agriculture and the Environmental Protection Agency at the Federal level and regional agencies at the local level. This intervention has not been homogenous as Federal and most regional agencies introduced stricter regulations, while Texas took the opposite route and further deregulated the sector (Bonanno and Constance, 2000). The following analysis speaks to this issue by reviewing the case of the penetration of corporate poultry production in rural regions of South-eastern Texas and the ensuing community-based resistance.

Sanderson Comes to Texas Farms

In 1995, Sanderson Farms, headquartered in Laurel, Mississippi, was the thirteenth largest broiler producer and processor in the US with US$393 million in sales (Feedstuffs, 1995; WATT Poultry, 2001). In early 1995, Sanderson Farms announced the end of the first phase of its expansion program. Phase One had increased its production 114 per cent from 1991. The second stage of expansion would begin in 1995 (Smith, 1995). Through the expansion strategy, by 2000 it had grown to the seventh largest poultry firm employing 8,147 people in its five integrated poultry complexes located in Mississippi, Louisiana, and Texas. Sanderson Farms markets over 500 different products shipped to every state in the US and many foreign countries (Sanderson, 1999a). On average the firm slaughters 5 million broilers per week. In the fiscal year 1999-2000, exports accounted for about 7 per cent of the total sales of US$603 million. The company's broilers are grown on contract with 495 farms in 2,075 houses with a capacity of 5.2 million placements per week (WATT Poultry, 2001).

In February 1995, Sanderson Farms announced that it was evaluating sites in Texas as the location for its next phase of growth. Chief Executive Officer Joe Sanderson Jr. said that the new complex would require a total investment of US$68 million for the new feed mill, hatchery and poultry processing plant, plus another US$56 million from the contract growers for pullet and broiler housing and growout equipment (Brown, 1995a). He stated, 'Texas offers a unique opportunity for the next phase of growth for Sanderson Farms' (Brown, 1995a, p.9).

Also in February 1995, Bryan, Texas, city officials met to consider a request from the Bryan-College Station Economic Development Corporation (EDC) to offer incentives to attract a new Sanderson Farms poultry processing plant to the area (Howell, 1995a). After Bryan-College officials visited a Sanderson Farms' processing plant in Mississippi, they commented that they were impressed with the company and its operations. John Anderson, President of the EDC stated, 'It would distress me if it went somewhere else. It's something that would be good for the community' (Howell, 1995b, A1). In a move to attract the Sanderson plant to the area, the Bryan City Council decided not to annex the area of the proposed plant for 15 years, a decision that would allow the plant to remain in Brazos County (Whitley, 1995). Soon after, the Brazos County Commissioners granted a ten-year tax abatement for Sanderson Farms. In defending the announcement, Commissioner Gary Norton stated, 'In today's way of doing business, if you don't play the corporate game you lose out' (Lambert, 1995, p.A1). County Judge Al Jones said that the court would continue to offer whatever incentives it could in order to compete with other areas. It appears that so long as there are entities competing for new business, Brazos County must participate in providing abatements, incentives and other benefits.

In May of 1995 Sanderson Farms announced that it would build its new processing plant in Brazos County and the new feed mill in adjacent Robertson County (Brown, 1995b). The new complex will be able to process 1.2 million broilers per week and will add 29 per cent to current production capacity (Meat and Poultry, 1995). Joe Sanderson, Jr. cited market accessibility to the 14 million people living within 200 miles of Bryan-College Station as the most important factor in the decision to locate in Brazos County. Other criteria in the site selection were local economic conditions and pro-business attitude, the residents' reception to the company and available resources such as utilities, potable water, and labour. According to CEO Sanderson, the new complex is designed to provide 'maximum marketing flexibility' through the production and processing of branded products, as well as value-added products, for both retail and food service market segments (Logan, 1995). At full staff, the operation will employ 1,400 workers, making it the largest non-governmental employer in the county (Taylor, 1997).

Sanderson Farms was promised more than US$600,000 by local governments, in addition to city and county tax abatements. It was also given 11.62 acres in the Bryan Business Park for its hatchery site. While some consumer and environmental activists opposed the location of the plant and the incentives provided to locate in Brazos County, supporters predicted that the city's investment in time and money would pay off. Opponents argued that the poultry litter contained arsenic that would contaminate local soils and water bodies. Other issues included concern over airborne stenches and the fact that the poultry processing jobs were low paying and would bring an influx of workers that would strain the social services of the community (Taylor, 1997).

The Sanderson Farms hatchery opened in October 1996, followed by the feed mill in January 1997, and the processing plant in March 1997. After opening a second processing line at the same location in July 1997, the plant reached its full first shift capacity of 650,000 birds per week in September. When double shifts of both lines are completed, the plant will reach its full capacity of 1.2 million birds per week. At full operation, the complex will support about 85 independent contract producers operating 48 breeder houses, 24 pullet houses and 320 broiler houses. The Texas complex provides Sanderson Farms access to the large and growing Texas market as well as allowing it to better serve customers in the Southwest and

West Coast (Sanderson Farms, 1999b). Bob Billingsley, director of development for Sanderson Farms, said the entire Brazos Valley community would benefit from the company. 'We feel it is our role as good corporate citizens to be a vital part of the community,' Billingsley said (Krinsky, 1998, p.E1). Texas Agriculture Commissioner Rick Perry commented, 'One of the very wise things Sanderson Farms did was work close with the community leaders before settling on Bryan-College Station. It will pay huge dividends for them.' (Krinsky, 1998, p.E12). In their 1999 Annual Report, Sanderson Farms states that 'one tradition that has been adhered to through the years, regardless of location, is a mutually beneficial relationship between Sanderson and the communities in which we work' (Sanderson Farms, 1999a, p.11).

From the Normangee Group to the Leon Country Environmental Group

While the processing plant is in the Brazos County Industrial Park, the hatchery is in the Bryan Business Park; the feed mill is in Franklin; the growout barns are spread through the countryside in Leon, Madison, Grimes, and Robertson Counties. Broiler growout barns are typically 40 ft. wide by 500 ft. long and house from 30-32 thousand chickens. Growout operations range from four to 26 barns on one site. The contractees mortgage their land to borrow money to construct the growout barns. Each barn costs about US$125,000. The average size operation of eight barns requires a US$1 million investment. The contractor arranges for the contractee to secure building loans. The contractor is responsible for delivering the day old baby chicks to the contractee, providing the feed and veterinary services, and picking up the birds for processing (after 40 days). The contractee is paid by the amount of weight the birds gain, a factor that is greatly impacted by the number of birds still alive at the end of the growing cycle. The contractee is responsible for housing costs, water, electricity, and labour (including picking up and incinerating the 'deads', ensuring that the watering and feed devices are working properly, and litter disposal).

Some of the first barns to go up were near Normangee and Flynn in Leon and Madison Counties. In the spring of 1998 '12 to 15 couples joined together to discuss the problems of the influx of chicken barns in the area' (Abernathy, 2000). Mr. Abernathy lives near Flynn and had already seen six broiler houses erected 'in his backyard'. The main concern of the group was to determine how many houses were going up in the area as they were already having problems with odour, flies, and increased respiratory problems. At the first meeting they decided to call themselves the 'Normangee Group'. Officers were elected and each family contributed US$100 for expenses such as mailings and newsletters (Abernathy, 2000).

Figure 7.1   Twelve chicken barns near Normangee, Texas

Source: Authors – picture taken from Texas State Highway

In June 1998 the Normangee Group retained a lawyer to try to stop the construction of some proposed chicken barns near Normangee. The attorney sent a formal letter to the contractee stating that his clients were concerned about the possible negative effects of the broiler barns on their lives and demanded that the contractees 'cease and desist all plans' to build the broiler barns. These effects included: foul odours that will drift across the neighbours property on a daily basis; a dramatic increase in the fly population due to the increase in chicken litter; dust from the feed and litter being blown on the neighbours' residences on a regular basis; and various noises such as cackling chickens, alarm bells and whistles installed in the poultry houses. In combination, the consequences of these effects 'will be that my clients will have to suffer substantial physical discomfort, annoyance and inconvenience in using their own homes, and the market value of their home and property will be diminished significantly' (Bennett, 1998a, p.1). Furthermore, the neighbours' health will be put at risk as a result of the long-term exposure to the barns. In response to this action, a

Sanderson Farms lawyer contacted the Normangee Group lawyer and informed him that the contractees were moving forward with construction and that 'Sanderson Farms would provide legal defence for this particular grower and anyone else who is sued for constructing houses' (Bennett, 1998b, p.1).

In July 1998, a representative of The Normangee Group wrote a letter to State Representative Steve Ogden complaining about the odours of the chicken houses in Leon County. The response letter acknowledged the request for investigation and turned the issue over to Texas Natural Resource Conservation Commission (TNRCC). A TNRCC representative was sent to Leon County to investigate the alleged odours and concluded that the growers were in compliance with the environmental regulations according to Texas standards. But the representative then stated that 'herein lies the challenge. The current odour regulations in Texas are subjective – no benchmark exists by which to measure the severity of or harm created by a particular odour...any new regulation must balance the legitimate and competing interest of two groups: private property owners and private citizens' (Ogden, 1998).

The Normangee Group held regular meetings, started a newsletter, and through word of mouth membership grew. The group decided that in order to include everyone in the Leon County area that they would change the name to the Leon County Environmental Group (LCEG). New officers were elected, members wrote letters to regional and national news programs and placed advertisements in local newspapers (LCEG, 1998a). At this time their attorney reported back to the group regarding the proposed lawsuit. Due to the unknown costs associated with a protracted lawsuit in which Sanderson Farms would provide legal support for its contractees, as well as the uncertainty of the potential outcomes, the LCEG Group reluctantly decided not to pursue the lawsuit (LCEG, 1998b). The LCEG renewed its letter writing campaign to state political representatives but all of the letters were forwarded to TNRCC. TNRCC responded to each letter and suggested that the group contact the Texas Department of Health regarding fly infestation concerns and local authorities for noise related issues (Saitas, 1998).

In late September 1998, the 'Poultry Summit' was held in Leon County to address the concerns of the LCEG. It was attended by State Senator Steve Ogden, County Commissioners, County Judges, local growers and representatives from the local State Representative's office, Sanderson Farms, the Texas State Soil and Water Conservation Board, the US Natural Resource Conservation Service, the Texas Poultry Federation, and the TNRCC, along with local and surrounding county news media. Issues discussed were the odour, flies, noise, property values, stockpiling of manure and the response time for complaints to the entities listed above. Even though all of these issues were addressed, nothing was resolved and Senator Ogden closed the meetings with comments on potential solutions such as limiting the number of houses within an area and having only one agency monitor all the problems (Johnson, 1998; The Normangee Star 1998).

Texans for Responsible Poultry Production

In January 1999, people from Madison and Grimes Counties increased their attendance at LCEG meetings to discuss the risks associated with poultry production in their areas. The LCEG members decided that to be inclusive and represent all Texans, they would change their name to Texans for Responsible Poultry Production (TRPP) (Abernathy, 2000). Each county would still be represented as a 'chapter' of TRPP. Andre Dean, a member of the Madison County chapter, commented that there were rumours of a chicken farm coming into their neighbourhood. According to Dean:

> We had heard of a bad situation with chicken farms in Leon County and some of us went to their meeting in December 1998. That was the spark to get going or we would suffer their fate of 12-24 chicken houses within 1 mile of this little community of 12 homes near Jewett that was under a nasty cloud of stench and flies, and valid concerns for the future of massive build-up of chicken manure with the real health concerns for our future. Concerns were stench, fly infestations, health concerns, lack of any enforceable controls or constraints by any government agency, inability of private citizens to sue for redress by state law protecting all agricultural projects from lawsuits after their first year of operation, manure run-off and loss of enjoyment of our own property as a violation of our 5th and 14th amendment rights to property (Dean, 2000).

Along with the new name change, new goals were set for TRPP. These goals were directed at changing legislation for poultry operations such as removing the AFO/CAFO distinctions for poultry producers (broiler operations were considered AFOs, animal feeding operations, and were not subject to the stricter regulations associated with CAFOs, confined animal feeding operations) and stronger waste management regulations (LCEG, 1999).

After several trips to visit State Senators and Representatives to express their concerns, TRPP members decided to draft their own legislation – the Enjoyment of Private Property Bill. They sent a draft of the bill to State

Senator Ogden and asked him to provide recommendations for improving the bill. Primary concerns addressed in the bill were:

- for the nuisance protection of poultry growers to be remanded;
- for nuisance to be defined by law;
- for the establishment of an odour tolerance zone;
- for mandatory setbacks from neighbouring properties; and
- that a permitting process be required for licensing of the broiler barns.

They also asked Senator Ogden to sponsor the bill in the Texas legislature (TRPP, 1999a). In April 1999, Senator Ogden and other State Legislators introduced a rider in the Appropriations Bill calling for US$100,000 in funding to conduct a statewide assessment of poultry operations. The rider provided funding for research to be done by the TNRCC and Texas A and M University on minimizing odour emissions and arsenic contamination from poultry facilities and finding the best method for taking corrective action when nuisance odours are confirmed (TRPP, 1999b). TRPP stated that this rider may help the growers, Sanderson Farms, and the neighbours come to a 'workable solution to what is considered a serious threat to the enjoyment of life and property...some of our most fundamental rights under the constitution' (TRPP, 1999b, p.1). In May the Enjoyment of Private Property Bill was hand delivered to two Senators and five Representatives prior to the meeting of legislature, along with over 1000 citizens' signatures. Although lauded by the legislators as an example of citizen democracy, the bill was not introduced for a vote of the legislature due to insufficient time to mobilize the necessary support. Senator Ogden assured the TRPP members that the utmost attention would be given to the matter of poultry operations and that the rider study would look at all sides of the issue, not just the side of Sanderson Farms, the growers, or the TRPP group (TRPP, 1999b).

The Study of the Neighbours to the Chicken Barns

In 1999 the Sociology Department at Sam Houston State University conducted a joint research project with TRPP investigating the community impacts of large-scale broiler production. Researchers attended the monthly meetings and worked with TRPP staff to develop a survey instrument to evaluate the views of the neighbours of the chicken barns regarding the social, environmental, and economic impacts of the chicken barns to address the concerns of the LCEG. A snowball technique was employed through providing selected TRPP members with 10-20 surveys for distribution. The convenience sample was geared toward those who lived in close proximity to the chicken barns. From the 160 surveys that were distributed, there were 63 usable responses creating a 39 per cent response rate (Constance, 2000).

The research reveals that almost all (93.2 per cent) the respondents have a sense of belonging and would be remorseful if they ultimately had to leave the community (95 per cent). About three-fourths of those surveyed indicated that the level of trust (73.7 per cent) and neighbourliness (75.9 per cent) has dropped in the community in the past few years. Overall, most respondents (82.8 per cent) reported that their community had become a less desirable place to live. Table 7.1 provides a sample of open-ended comments regarding what factors had led to the perception of decreased quality of life since 1997. Respondents blamed the recent introduction of the chicken barns and the associated negative effects on their lives. Commonly reported effects include odours, flies, land/water contamination, health problems, and decreased property values (see Table 7.1).

Regarding the perceived impacts that the livestock odours have on the quality of their lives, respondents indicated that because of the odours they had restricted outdoor activities (76.2 per cent), kept windows closed (80.5 per cent), had fewer visitors (56.4 per cent), and had experienced increased tensions with neighbours (81.4 per cent). Furthermore, most people had concerns about their family's health (90.0 per cent) and many reported respiratory problems (64.9 per cent) and burning of the eyes, nose, or throat (56.4 per cent). About one third (34.3 per cent) of those surveyed indicated that they had lost friends due to the recent advent of the poultry barns.

When asked about the impacts of the chicken barns, respondents reported mostly negative impacts on their property rights/values (90.0 per cent), personal and family health (86.6 per cent), and community neighbourliness (86.4 per cent). Similarly, a large majority reported negative environmental (89.9 per cent) and social impacts (84.5 per cent), as well as a general decrease in the overall quality of life (89.6 per cent) due to the chicken operations. Over 90 per cent of respondents felt that the level of regulation (93.1 per cent) and enforcement (94.9 per cent) of the poultry industry was not adequate. The vast majority (77.2 per cent) also felt that local government should be able to regulate the poultry industry. When asked about increased levels of regulation, while almost all people (98.2 per cent) felt that the poultry industry should be more regulated, only about one-half (53.7 per cent) felt that it would be. Overall, about 80 per cent of respondents disagreed that the benefits of the recent introduction of the chicken industry outweighed the costs.

Table 7.1   Responses to the question: what exactly has changed to make your community more or less desirable than in 1997?

Fighting over the chicken houses.
1) Personal health concerns. 2) Property value is decreasing – chicken producers have made this a less desirable place to live. 3) The air stinks. 4) Fly problems increased. 5) Arsenic in manure of chickens is spread on neighbouring fields causing run-off onto our land – polluting our soil with arsenic. We have increased concern about the quality of our drinking water and the quality of our lake water. Personal depression after working all your life. IT IS NOT RIGHT!
Less desirable: Nextdoor neighbour has become an 'indentured servant' for Sanderson Farms; has built 8 chicken houses (32,000 birds each) and is a major threat to health by water and air pollution. In the near future will (arsenic poop and dead chickens) contaminate water table? Ground water (7 acre lake) is immediate threat. Land value is devalued. We are following his (neighbour to north) agenda and life style, not our own. HELP!
You know that the smell of the chickens has influenced my opinion of quality of life here. Going outside and being overwhelmed by someone else's making money is really hard on a lady 75 years old!
Two commercial chicken growers have established houses in the immediate area causing residents to suffer physically and financially. We have a feeling of being trespassed upon and no laws to protect us.
I cannot open the windows in my home. I have to plan church group activities, barbecues and parties around the chicken houses, and when the smell will be the worst. I cannot go out to the pasture to check the cows on certain days without contracting a headache and nausea. The fly population is growing inside my house. My asthma has worsened.
My husband and I built a house six years ago. Now we have chicken houses half a mile down the road. The smell is awful. My children are always sick and my property value has decreased dramatically because of the chicken houses. Chicken houses have ruined our community, friendships, and in my opinion, are a health hazard. If I had the money, I would move.

Citizens Against Poultry Production

In their first meeting of 2000, TRPP asked political candidates to attend and discuss their positions on the growth of the poultry industry in the region. Most candidates said they were not aware of the extent of the growth of the industry in the 'out counties' and expressed their 'concerned about the proximity of the chicken houses to populated areas and about property devaluation' (TRPP, 2000, p.1). At this meeting an executive board was elected including one representative each from Leon, Madison, Robertson, and Grimes Counties and one for the North Texas Chapter and the East Texas Chapter. The Chairpersons of each group met each month and report back to their respective groups. The entire group met every three months. It was also decided to change the name of the group to Citizens Against Poultry Pollution (CAPP) to remedy a perceived confusion regarding the meaning of TRPP (TRPP, 2000). The CAPP members renewed their commitment to use the legislative process to change state laws related to the regulation of the poultry industry and the protection of their private property rights (CAPP, 2000a).

In the summer of 2000 the results of the research funded by the rider were announced. CAPP was very disappointed with the results of the study done by the TNRCC and Texas A and M University. CAPP felt that the study report was only an extensive literature review and did not contain a substantive research dimension (CAPP, 2000b). CAPP argued that while the TNRCC had done some air testing at one site in Madison County, the testing was useless because there were no chickens at the facility when the testing was conducted (CAPP, 2000c). At this time, CAPP continues to pursue its legislative agenda.

Globalization, Sustainable Livelihood, and Resistance

Globalization

The events of the case presented above speak directly to the issue of globalization and its consequences to local communities. As clearly illustrated in pertinent literature (e.g. Bonanno and Constance, 1996; Gray, 1998; Boggs, 2000) globalization emerged as a political project to counter the limits experienced by Fordism. Under the most advanced form of Fordism (circa 1945-1975), socio-economic development was successfully managed by an interventionist nation-state that controlled unwanted consequences of capitalism through enhanced planning and coordination.

The Fordist nation-state and the related international state system developed inclusionary policies which allowed the – albeit partial and often contradictory – inclusion of subordinate groups into mainstream society. This *labour-capital accord* allowed fast and sustained capital accumulation in manufacturing and agriculture, but also contemplated a number of important concessions to working and subordinate classes. The growth and strengthening of the welfare-oriented policies that characterized post-WWII development capture the mood and culture of the Fordist regime.

Fordism was functioning at nearly optimum levels from the middle 1950s to the late 1960s. But in the early 1970s the regime was faltering badly and exposing serious contradictions. The rise of new social movements, student protest, Third World anti-western resistance and economic downturns eroded the Fordist stability and allowed critics to characterize state interventionism as the cause of the economic contraction and the source of crippling rigidities. New strategies aimed at weakening the power of organized labour, promoting free market policies, and slashing social programs were introduced with wide support. Critics held that Fordism produced a crippling profit-squeeze that endangered capitalism and that a substantial part of the Fordist institutional and ideological complex ought to be dismantled. Most importantly, many of the core Fordist policies which expanded opportunity and rights were made prime targets of strategies aimed at increasing the freedom of property holders at the expense of wage workers and subaltern strata, and distributing wealth and power upward. A pervasive sense that post-war structures of accumulation failed and that social policy needed to be reversed quickly emerged. Efforts to reignite growth weakened or eliminated post-war mechanisms aimed at increasing equal opportunity, providing for the unemployed and needy, and blocking the colonization of valued non-economic environments by capital. Rather than equal rights, the costs of regulation and their need to increase discipline and security became top priorities.

This globalization project promised emancipation through its primary feature: de-regulation. It meant a frontal attack against the interventionist role of the nation-state in favour of the implementation of a free-market based re-regulation project. The policies governing socio-economic development were rewritten to enhance the free mobility of capital and diminish the means through which communities and governments once controlled unwanted consequences of capitalism. This project's underling assumption was that the free-market oriented policies offered more efficient solutions to socio-economic problems than their now exhausted Fordist antecedents.

Sustainable Livelihood

The globalization project received support from a variety of groups. While advocates of neo-liberal positions enthusiastically endorsed de-regulation and re-regulation projects (Friedman, 2000), even progressive oriented groups – which criticized Fordism for its too oppressive regulatory component – showed support for the new global regime (Hirst and Zeitlan, 1991). Texas emerged as one region in which the convergence of these two dimensions is clearly visible. For the past two decades the improvement of the socio-economic conditions of Texas residents has been promoted through free-market oriented policies, de-regulation and the fostering of a pro-free market cultural climate (Bonanno and Constance, 2000). It is not a coincidence that for the past two local general elections, key state posts were won by Republican candidates who supported pro free-market platforms in which the attraction of investments in unregulated and/or minimally regulated sectors was equated with socio-economic expansion for everyone. Doing business in Texas meant to 'play the corporate game' because for local communities this action 'pays good dividends'. This was the socio-cultural terrain that attracted Sanderson Farms to Texas as the base of its 'second phase of expansion' to service the global agro-food market in a more flexible manner. In this respect Sanderson Farms acted following a scenario that has been often promoted and practised under globalization – global sourcing.

The events of the case reveal the problematic nature of this project. Sanderson Farms benefited from the free-market, pro-business climate of the region, the strategic location of its production facilities, and the human and natural resources available. These are rural areas which experienced a constant decline of their local traditional agricultural operations (mostly cattle ranching), were located in close proximity to a very large metropolitan area (Houston) and a gateway to even larger markets and had no regulation concerning the location of intensive poultry production facilities. In a global sourcing framework, these counties presented clear advantages over competing areas. Additionally, growers supported Sanderson Farms' initiative. They viewed the poultry industry as an opportunity to renew economic viability despite the onerous and risky initial investments. Corporate claims of community involvements notwithstanding, invitations to participate were extended with discretion, planning was confined to exclusively the poultry processing plant in Bryant, and only a few people in the four county area knew of the new initiative before the barns were actually erected.

As in other episodes of globalization, Sanderson Farms' project operated well at the economic level but failed at the legitimative level (Dicken, 1998; Gray, 1998). This corporate project did not contemplate tangible community involvement and, while receiving cooperation from the upper strata of the State of Texas, equated the absence of formal requirements to consult with local residents with substantive community participation. Following dominant globalization project and pro-free market rhetoric, Sanderson Farms operated as if business expansion would automatically translate into community wellbeing and did not plan actions directed at addressing undesirable consequences of economic expansion. The net result was that while corporate expansion continued, the communities involved did not benefit. This is a novel, yet now typical, condition of globalization-based economic projects: economic expansion is decoupled from community socio-economic development.

Resistance

The decoupling of economic expansion and community socio-economic wellbeing and the absence of legitimizing mechanisms to address emerging problems served as vectors to radicalize community response to corporate penetration. The extent of the local response is arguably the most surprising aspect of the situation as it represents a departure from early readings of global sourcing. In these previous analyses, corporate hyper-mobility and local underdevelopment were viewed as factors that either suppressed resistance (McMichael, 1996) or left room for only community-based redefinitions of situations (Arce, 1997). In the present case study, resistance was engendered by the false unity between claims of socio-economic wellbeing and the degraded reality of community living. While environmental degradation was a primary factor in the mobilization of local resistance, it was the issue of property that served as the catalyst for resistance. It was the inability of this corporate project to provide legitimate answers to the loss of property value experienced by community residents that triggered the widest opposition. In a three-year span, the resistance group grew from 12 families to a statewide organization with an active legislative agenda and political strategy. In this context, resistance was fuelled by the inadequacy of the corporate free-market discourse to address individual property value issues in a context in which they were at stake and ultimately devalued. It is relevant to note that both free market and property value are items that often co-exist in free-market oriented discourses.

Traditionally, legitimation issues have been addressed through the intervention of the state (Offe, 1996). Under the globalization project, however, state powers have been either greatly reduced and/or transformed to accommodate capital mobility and requirements preventing effective mediation. Following the globalization project, the officialdom of the state of Texas' intervention was heavily tilted toward the protection of corporate interests. The assistance the state provided to Sanderson Farms, the lack of support for regulation of AFOs, and the inadequacy of the state promoted scientific study of the issues at hand are all indications of the limits that the current form of the local state encounters in mediating conflicting demands. This is perhaps the most important conclusion to be reached from the analysis of this case: the opening of local communities to globalization lacks institutions capable of buffering the unwanted consequences of the growth of capitalism and controlling its most powerful actors. This situation engenders fierce resistance and creates a contested terrain in which corporate power is resisted and redefined even in conditions in which the demand of local residents are grounded in discourses parallel to those of corporations.

References

Abernathy, T. (2000), Personal interview, 5 May.

Arce, A. (1997), 'Globalization and Food Objects', *International Journal of Sociology of Agriculture and Food,* vol. 6, pp. 77-107.

Bennett, W. Jr. (1998a), Normangee Group lawsuit letter, 11 June.

Bennett, W. Jr. (1998b), Memorandum No. 3 to Normangee Group.

Boggs, C. (2000), *The End of Politics. Corporate Power and the Decline of the Public Sphere,* The Guilford Press, New York.

Bonanno, A. and Constance, D. (1996), *Caught in the Net. The Global Tuna Industry, Environmentalism and the State,* University Press of Kansas, Lawrence.

Bonanno, A. and Constance, D. (2000), 'Mega Hog Farms in the Texas Panhandle Region: Corporate Actions and Local Resistance', *Research in Social Movements, Conflicts and Change,* vol. 22, pp. 83-110.

Brown, R. (1995a), 'Sanderson Farms Expanding with Southwest Complex', *Feedstuffs,* vol. 67(10), p. 9.

Brown, R. (1995b), 'Sanderson Farm Building Poultry Complex', *Feedstuffs,* vol. 67(22), p. 7.

CAPP (Citizens Against Poultry Pollution)(2000a), Monthly Newsletter, 23 March.

CAPP (2000b), Monthly Newsletter, 27 September.

CAPP (2000c), Monthly Newsletter, 11 November.

Constance, D. (2000), 'Final Report: The Community Impacts of Large-Scale Chicken Production East Texas', 29 April.

Dean, A. (2000), Personal interview, 8 May.

Dicken, P. (1998), *Global Shift,* The Guilford Press, New York.

Feedstuffs (1995), 1995 Reference Issue, 19 July, p. 21.

Friedman, T. (2000), *The Lexus and the Olive Tree,* Anchor Books, New York.

Gray, J. (1998), *False Dawn: The Delusions of Global Capitalism,* The New Press, New York.

Hirst, P. and Zeitlan, J. (1991), 'Flexible Specialization Versus Post-Fordism: Theory, Evidence and Policy Implications', *Economy and Society,* vol. 20(1), pp. 1-56.

Howell, D. (1995a), 'Mississippi Plant Looking at Brazos Valley', *The Eagle,* 17 February, p. A4.

Howell, D. (1995b), 'Lucky Pluck: B-CS Still in Hunt for Chicken Plant', *The Eagle,* 23 February, p. A1.

Johnson, S. (1998), Letter to attendees of 23 November poultry meeting, 23 December.

Krinsky, M. (1998), 'Winging It: Sanderson Farms Chicken Operation Prepares to Lift Off', *The Eagle,* 2 September, p. E1, p. E12.

Lambert, C. (1995), 'Chicken Plant Deal Spreads Its Wings', *The Eagle,* 15 December, p. A1.

LCEG (Leon County Environmental Group)(1998a), Meeting minutes, 16 August.

LCEG (1998b), Meeting minutes, 9 July.

LCEG (1999), Monthly Newsletter, 27 January.

Logan, S. (1995), 'Sanderson Farms Breaks Bryan Ground', *The Eagle,* 2 October, p. A1.

McMichael, P. (1996), *Development and Social Change.* Pine Forge Press, Thousand Oaks.

Meat and Poultry (1995), *Companies,* vol. 41(11), p. 8.

Offe, C. (1996), *Modernity and the State: East, West,* MIT Press, Cambridge.

Ogden, S. (1998), Letter to Normangee Group member Calvin Hodde, 15 July.

Saitas, J. (1998), Letter to US Senator Kay Bailey Hutchinson, 1 October.

Sanderson Farms (1999a), Sanderson Farms Building Tradition: 1999 Annual Report

Sanderson Farms (1999b) Sanderson Farms: Simply the Best, 18 May 1999, http:///www.sandersonfarms.com/info/where.html.

Smith, R. (1995), 'Sanderson Farms Expands 114 Per cent in First Phase of Growth Strategy', *Feedstuffs,* vol. 67(8), p. 6.

Taylor, M. (1997), 'Here Comes Chicken Big', *Insight,* January, pp. 27-33.

The Normangee Star (1998), *Environmental Group Gets Little Help When Big Guns Show up for Meeting,* vol. 89(39), p. 1.

TRPP (Texans for Responsible Poultry Production)(1999a), Monthly Newsletter, 12 April.

TRPP (1999b), Monthly Newsletter, 2 May.

TRPP (2000), Monthly Newsletter, 22 February.

WATT Poultry USA (2001), *Nation's Top Broiler Companies,* vol 2 (21), p. 26d.

Whitley, C. (1995), 'Sanderson Farms One Step Closer After Council Meeting', *The Eagle,* 3 April, p. A3.

Chapter 8

Globalization and Agro-Food Systems in Argentina

Miguel Teubal and Javier Rodríguez

Introduction

Since the mid-1970s important transformations have taken place in Argentina's agro-food system (AFS), that is the socio-economic space that includes agriculture, the provision of agricultural inputs and the marketing, processing and final distribution of food. AFS is the totality of these relations 'as they affect the production of food and access to it' (Barraclough, 1982, p.125).

These transformations can be related to:

- Global macroeconomic policies – the adjustment policies applied by successive governments, in particular the structural adjustments (privatizations, deregulations and opening up to the world economy) of the Menem Administration (1989-1999);
- Sectoral policies influencing agriculture and food, specially the so-called Decree 2284 of Deregulation of November 1991 whereby a series of activities were deregulated including many having to do with the production, processing and marketing of food and agricultural produce. Amongst other deregulatory measures, all of the existing agricultural boards, some of them created in the 1930s, were dissolved. These included the Junta Nacional de Carnes (National Meat Board), Junta Nacional de Granos (National Grain Board), and the Dirección Nacional del Azúcar (National Sugar Directorate);
- These measures lead the way to the introduction of a series of technological and organizational changes and power relations in the different sectors and complexes that form the AFS induced, in large measure, by globalization processes and structural adjustment policies to which we make reference above.

Thus, wholesale privatization, deregulation, and liberalization of foreign trade and investment policies introduced in the 1990s opened the way to multiple globalization processes. Large agro-food corporations, including transnational or transnationalized supermarkets, increased their leverage in Argentina's AFS. The use of new inputs and technologies controlled by these large corporations was intensified. Concentration and centralization of capital and vertical integration in each and every one of the complexes that form the Argentine AFS was intensified. And, toward the end of the decade, large foreign agribusiness took over much of Argentina's food system through joint ventures or purchases of local companies. Thus, in the 1990s an AFS, traditionally comprised of a series of medium and small sized companies that coexisted with large local agribusiness, changed its structure and profile. Concentration and centralization of capital, greater influence of foreign companies and increased vertical integration all led to changes in the power relations within, and technological profiles of, the Argentine AFS. All these factors influencing articulations within the agro-food system affected medium and small farms, urban and rural workers of the agro-food system, as well as access to food for vast social sectors. Contract farming and other forms of agro-industrial articulation contributed to increased vertical integration, increasing the power of large extra-agrarian corporations vis-à-vis medium and small farmers that significantly lost their decision-making autonomy.

How can we characterize the recent transformations that took place in Argentina's AFS? What are some of their consequences for medium and small farmers, rural and urban workers of the food system, and medium and small business operating in the food industry and in the final distribution of food? Given the advance of large corporations in the food industry and the thrust of supermarkets, what were the impacts on 'access to food' and on the quality and nature of food for vast segments of the population?

In this chapter we analyse some of these processes. In the first place we present an overall vision of the Argentine agro-food system, the complexes that integrate it and some of their characteristics. Then we highlight some of the transformations occurring in recent decades, and point out their consequences for the diverse social agents that form the Argentine food system.

Agro-food Complexes in Argentina

The Argentine agro-food system can be subdivided schematically into 14 agro-industrial complexes of varied importance and characteristics –

including rice, tea and *yerba mate*, poultry, fish, sugar cane, wine, tobacco, fruit and vegetables, milk, meat, oilseeds and grain. These complexes range from US$200 million in production per year for rice to over US$4,500 million for grain (INDEC, 1995). All include agriculture and agro-industrial production as well as retail and wholesale final distribution and consumption of food – oriented either to the domestic or overseas markets.

The grain, oilseeds, meat and milk complexes based mainly in the Pampaen Region, are the main complexes in terms of their contribution to national product. The first three are also export complexes that define Argentina as a 'new agriculture country' (Friedmann, 1992). The grain and meat complexes are also important generators of employment, followed by the milk, and fruits and vegetables, complexes. The poultry and milk complexes grew significantly in recent years – the latter on the basis of exports to Brazil. Mention can also be made of the wine, sugar cane, tea and *yerba mate* complexes producing the traditional 'industrial crops' of the interior that have an important regional significance.

Recent Transformations

In recent years most of the agro-food complexes of Argentina have been considerably transformed and their multiple internal articulations have been modified. In general, concentration and centralization of capital increased in food manufacture and in the final distribution of food (the latter having to do with the growth of supermarkets). The formation of conglomerates and the increased importance of foreign companies and large transnational agribusiness needs also to be mentioned. These tendencies are related to one other – a greater influence of foreign companies contributed to concentration and centralization of capital. Both these factors led to increased vertical integration. Associated with these processes was the introduction of new technologies that contributed to increasing asymmetries within the complexes being considered. All these processes impacted on urban and rural employment, the farm sector as a whole and the role of rural society in the overall agro-food system.

Of course these transformations operated variously in different complexes, being on the whole observable in each and every one of them.

Concentration and Centralization of Capital in Industry, the Formation of Conglomerates

Industrial concentration increased considerably in the last decade, due to the varied managerial strategies adopted by large firms that tended to increase or maintain their shares in different markets. In a climate of extreme deregulation, concentration permitted large companies to strengthen their negotiating power vis-à-vis their suppliers, including farmers. Increased concentration was also reflected in exports where a few companies control an increasing share of total exports. Examples of large economic groups absorbing a series of companies thus increasing their participation in different markets can also be mentioned.

The Industrial Census of 1994 elaborated by the National Institute of Statistics and Census (Instituto Nacional de Estadística y Censos, INDEC) gives us a notion of the degree of concentration existing in different industrial branches. (Census data considers premises, 'locales' in Spanish, as the unit of analysis. It is evident that many firms own a series of industrial premises, implying a larger degree of concentration then what is here presented.) These data permit us to observe a wide diversity in the degree of concentration among different complexes and within each one. For example, in the grain complex, highly concentrated activities coexist side-by-side with activities that are not as concentrated – in the 'milling of starch, and of vegetables and cereals other than wheat', the first eight premises of this industrial branch produced 84.1 per cent of the total production; on the other hand in the 'processing of bakery products' the eight most important premises generated only 17.4 per cent of total production.

This great diversity in the degree of industrial concentration denotes the fact that companies operating in key stages of the agro-industrial chain can influence the rest of the complex by establishing purchasing conditions, prices, or rules of different sorts.

The tobacco complex is one of the more highly concentrated complexes with the eight largest premises producing 99.5 per cent of total production. This is followed by the brewer complex where eight firms produce 91.2 per cent of industrial production. In the rice complex the corresponding proportion amounts to 65.5 per cent, in the sugar cane industry 87.3 per cent and in *yerba mate* 81.8 per cent. The main oilseeds firms produce 61.2 per cent of production. Much less concentrated is the meatpacking industry: the eight most important firms produced 26.1 per cent of the total production, reflecting on the whole, the persistence of a large quantity of small and medium industries in this complex. The low degree of concentration in this complex is also reflected by the fact that sales of its main firms (Swift and CEPA) have a lower value than the sales of the main companies of other complexes, for example, Sancor and La Serenísima in the milk complex, or Arcor and Pepsi and Coca Cola in the sugar complex.

But beyond the diversity in industrial concentration as among different complexes and within each complex as was noted in the grain complex, concentration levels have increased over time when comparisons are made with the previous census of 1985. What can be shown on the basis of different sectoral studies is that in the period 1985-1994 there was an increase in industrial concentration in almost all of the complexes being considered.

Concentration and Centralization of Capital in Final Distribution. The Thrust of Supermarkets (Supermaketization)

For many decades the retail distribution of food in Argentina was through a large number of small shops. Wholesale distribution was also relatively competitive. In the early 1970s more than 80 per cent of food sales to the public was channelled through these traditional and specialized small shops: groceries, bakeries, cold meat, butcher and vegetable and fruit stores. In that decade some local capital supermarkets were established and retail distribution began to change. In the 1980s large foreign capital hyper- and supermarkets – Carrefour, Jumbo and Makro – made their entrance into Argentina's AFS. In the 1990s, the massive diffusion of supermarkets took place. In less than 20 years, retail distribution changed totally its profile. This had important consequences for the agro-food system as a whole.

Supermarkets appear as a new social actor that embraces all the complexes in their final stage. The remarkable growth of supermarkets is an essential factor influencing changes in the structure of each complex.

As is shown in Figure 8.1, the supermarkets' share of total sales doubled in almost 10 years, to the detriment of traditional retailers. While in 1984 in the Greater Buenos Aires Area supermarkets absorbed 24 per cent of retail sales of mass consumption goods, mainly staple foods, by 1993 this had risen to more than 50 per cent. One third of those sales were made by the eight most important supermarket chains (Carrefour, Coto, Norte, Disco, Tía, Jumbo, Macro and Su Supermarket). This increased *supermaketization* contributed in large measure to increasing the degree of concentration operating in the food system.

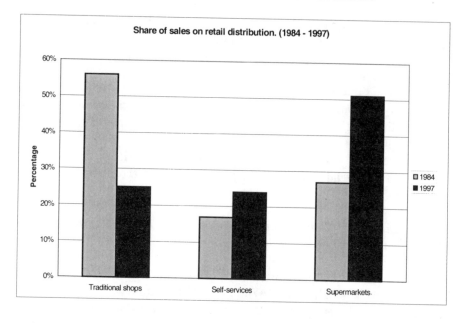

Figure 8.1   Share of sales on retail distribution

Source: Sagpya based on data of the Cámara de Supermercados (Chamber of
Supermarkets)

Some of the main supermarkets thus had higher sales than some of the
main food industries processing traditional foods – a factor that allowed
them to substantially increase their negotiating capacity. In the context of a
complete deregulation of the economy, supermarkets were able to impose
upon their suppliers a series of conditions – postponing payments,
obtaining special allowances, and so on – that increased substantially their
profit margins.

The spectacular and sustained growth of supermarkets was followed by
a series of mergers; for example that of Carrefour and Promodes on the
international level, and the local Tía and Norte merger. In Argentina the
Carrefour with Promodes merger implied the association of Carrefour with
the previous Norte-Tía merger.

Many supermarkets were sold to transnational companies; others were
absorbed by large local capital companies. Concentration and centralization
of capital in the retail sector decreased the negotiating capacity of other
social actors within each complex. These included consumers and farmers
that increasingly visualized their options as being restricted by the
increasing oligopolization of the system.

Take-overs by Foreign Companies

Together with processes of concentration and centralization of capital, in the last decade there was a notable advance of take-overs by foreign companies – usually large transnational agribusiness of the main Argentine food companies. A survey by INDEC (INDEC, 1999a) clearly highlights this process. This survey also presents information concerning the insertion of some transnational firms in the local scene, substantially transforming these complexes.

When considering the largest 500 firms of Argentina, in less than five years the share of foreign companies in the 'food, beverages and tobacco' branch grew by 25 per cent. This increase of the participation of foreign firms in the Argentine food industry is similar to that occurring in the rest of the economy. However, it should be pointed out that the share of foreign capital in the food industry is comparatively lower than the average of the economy as a whole. This situation continues even today, despite the fact that the local landing of foreign companies was stronger in this sector on average than in the rest of the economy.

Vertical Integration

As mentioned above, in the last decade vertical integration increased in each of the complexes being considered. Indeed, a policy frequently used by large economic groups consisted of the acquisition of companies or the development of new ones that allowed vertical integration in a food chain. This strategy helped to determine what, how and with what technologies production in agriculture and in the other stages of the complex should be carried out with greater efficiency.

According to Heffernan vertical integration is a strategy of monopoly capital (Heffernan, 1998). It occurs when a company, generally a manufacturing industry or supermarket, controls several stages of the productive process of a certain product, or group of products. While horizontal concentration contributes to the accumulation of economic power of large companies in certain stages and geographical regions of the food chain, vertical integration exercises the same effect, by closely articulating the diverse stages of the transformation process:

> Although there are great variations among the different commodity sectors regarding the ways concentration of ownership and control have occurred in the processing and distribution stages, the same general pattern of fewer and larger firms in each stage has been underway during the last half of the twentieth

century in the United States and has become most obvious during the last decade at the global level (Heffernan, 1998, p.49).

While horizontal integration usually refers to expansion in the same stage of the same commodity sector, vertical integration occurs when a firm increases ownership and control of a number of stages in a commodity system or agro-industrial complex.

In general there can be different degrees of vertical integration. When the degree of vertical integration of a complex is very high, we can assume that the articulations within this commodity system are intense and that there exists a high degree of coordination with regards to the decisions taken by a nucleus of power, usually a food industry or supermarket, and those taken in another parts of the complex (Vigorito, 1979; Teubal, 1999). Thus, if a company owns the land on which the raw material is produced, the industrial plant that processes it and the supermarket that distributes the food in its final form, full vertical integration exists: the agricultural sector operates as if it was a 'supply department' of the company, and agriculture production, industrial processing and final distribution can be very closely coordinated.

In each agro-industrial complex or commodity system, diverse institutional mechanisms of vertical integration exist articulating different parts of the whole. Vertical integration on the basis of property rights is only one form of vertical integration. Another is contract farming. This occurs when a large company articulates with independent farmers by means of contracts that can take on varied forms. Even if this institutional form of articulation does not exist, vertical integration may increase due to the greater market power that the large company can exercise, as a consequence of greater monopolization or oligopolization of production and/or distribution. It would seem that the increase of monopsonic or oligopsonic power of the large manufacturing companies or supermarkets also increases vertical integration in its different forms.

Vertical Integration Based on Ownership or Property Rights

Vertical integration based on ownership and control occurs when a company or economic group integrates legally, as part of a single property unit, diverse economic activities corresponding to two or more stages of an agro-industrial complex. Vertical integration is full-fledged when it embraces all the stages of a commodity system, from the production of agriculture commodities, their processing and transformation, to the final distribution of food. Examples of vertical integration by ownership and control in Argentina are the cases of the companies Arcor and Cargill.

The Arcor group operates mainly in the sugar cane complex, having begun its activities as a producer of candy and other sweets. Later, it also intervened in other commodity systems – in oilseeds (Arcor produces vegetables oils) and in the fruit and vegetable complex (producing preserves and mashed tomatoes).

A characteristic of this group that grew notably in the 1990s is that of the vertical integration of production. Arcor not only manufactures different commodities, but also produces packaging products and the material with which this packaging is made. In fact it is the main producer in Argentina of corrugated cardboard. In its various industrial plants Arcor manufactures crackers, cookies, puddings, candies, sweets and marmalades, fruit pulp, flavours and essences, powdered juices, *alfajores* (a kind of pastry made in Argentina) sweet bread, tomato preserves and mashed tomatoes, oils, glucose and so forth. The Arcor group includes the companies Metalbox SA and Frutos de Cuyo SA that produce tin containers, Converflex and Flexiprim producing flexible containers, Industrias Mark that manufactures filler machines, and Cartocor SA corrugated cardboard and boxes. The group also owns a factory that produces wrapping film.

In 1996, Arcor acquired the sugar mill Providencia in Tucumán. Sugar is one of the main inputs of its food industries. This sugar mill developed contract farming with the local sugar cane producers – more or less guaranteeing the supply of the main input, not only for this sugar mill but also for the rest of the industries of the sugar complex pertaining to Arcor.

Vertical integration of the Arcor group is not only *upstream,* that is with agriculture, but also extends to the distribution of sugar and its products. In 1997 Arcor bought Bertolo a company that distributes its products in the Province of Córdoba. In the Province of Buenos Aires, Dicami S – another member of the group – distributes Arcor products.

Another example worth considering is the case of Cargill. Traditionally, Cargill in Argentina operated in the marketing of grains. However, the strategy of this company has gone beyond this activity. Being one of the main processors of oilseeds in the world, Cargill expanded its activities in the local vegetable oil industry becoming one of the largest Argentine exporters of vegetable oil and protein flour and their subproducts. The participation of Cargill in local industry increased in the 1980s when oilseed plants were established in Quequén and Bahía Blanca (Province of Buenos Aires) and Puerto San Martin (Province of Santa Fé).

In the 1990s this company's vertical integration strategy become more visible when it purchased Minetti, a flour mill and noodles factory, as well as the purchase of several grain mills at the beginning of 1995. As a consequence, Cargill became an important producer and trader of flour and other grain products. The vertical integration that was thus achieved, together with the greater facilities this company has for access to credit as a multinational group, were used to secure larger portions of the market.

Vertical Integration by Contract

In this case, vertical integration is carried out via contract farming. A contract generally establishes the quantities and prices of the raw materials the industry is willing to pay, but also other conditions regarding quality of production, periods of delivery and so forth. Some contracts are more thorough than others, relating to more than one stage of production within a commodity system (Teubal and Pastore, 1995). This form of integration is found in the fruit and vegetable, sugar cane, milk, poultry and brewer complexes among others. While many examples can been presented (see Teubal and Pastore, 1995), here we consider the recent case of contract farming in the potato industry.

The recent consolidation of the local fast-food industry implied an increase in vertical integration, since this food chain has a policy of controlling earlier stages of production. The fast food firms not only control the industry but also influence potato farmers. The industry generates contracts with the potato farmers whereby certain prices and qualities for certain quantities of purchases are established apart form other purchasing conditions.

The quantity of potatoes produced under contract has grown in recent years reaching 20 per cent of total production in 1998 (Mateos and Capezio, 1999). Besides specifying certain quality demands in the contracts, there is a requirement peculiar to the industry: the use of certain specific potato varieties. Indeed, for conveniences associated to industrial yields, the varieties Shepody and Russet Burbank were introduced, displacing the traditional Kennebec variety. It can be pointed out that the Russet Burbank potato variety is not good for the direct fresh consumers' market because of its rough skin, making industry its sole destination. It is clear that this potato variety creates a still greater dependence on the industry on the part of the farmer. It shows clearly how industry tends to become an integrative pole in this commodity system.

Concentration and Vertical Integration: The Milk Complex

Concentration in the milk industry led to a decrease in the total number of dairy farms – a factor that clearly shows how firms operating in a later stage of the productive process can influence those operating in previous stages, either through a policy of differentiation or of cooperation and selective integration.

According to the National Economic Census of 1994, at that time there were 738 milk processing establishments in Argentina compared with the 1892 registered in the previous census of 1985. Nevertheless, 22 companies processed 73 per cent of all milk produced. The three largest companies are: Sancor (a second degree Cooperative), Mastellone (La Serenísima) and the transnational Nestlé. Each of these firms is a leader in the production of different milk products – Sancor in cheese and butter, Mastellone in fluid milk and Nestlé in powdered and condensed milk.

Industrial concentration led to vertical integration, and to a weakening of the negotiating capacity of the small independent milk farmers. A highly concentrated industry that operates with large volumes of milk tends to impose a series of conditions on the small milk farmers. We should notice here that industrial concentration is much larger when geographical regions are considered. It is not easy for a milk farmer to find another industry apart from the local one that can buy the product.

Technological Changes and Biotechnologies: Their Contribution to Increased Vertical Integration

Part of these large transformations in the last decade is related to the introduction of new technologies, in particular, biotechnologies. It is possible to highlight the introduction of transgenic seeds. Since 1996 when CONABIA (the National Commission of Biotechnologies) approved the use of the Roundup Ready (RR) soybean in Argentina, its use expanded rapidly. Indeed, while in the 1996/97 campaign only about 100,000 hectares were planted with transgenic soya beans, the following year 1,400,000 hectares were used to produce transgenic soya beans. At present 85 per cent of the soybean production in Argentina is transgenic.

The phenomenal diffusion of the transgenic soybean can be explained by the reduction in the weeding costs as well as other planting and harvesting costs. A certain weakness in the legal context can also be one of the reasons, since legally no objection could be made to the diffusion of genetically modified organisms.

With regard to the transgenic soya, the RR soybean tolerates the herbicide glyphosate that must be used as a necessary complement to the acquisition of the transgenic seeds. Monsanto is the only producer of glyphosate in its herbicide 'Roundup'. Thus, once farmers incorporate the transgenic seed into production they are forced to buy the glyphosate that only Monsanto produces, at a price that the grower cannot negotiate. It is clear that farmers have become more subordinate than before to the supplier of seeds and herbicides, with the introduction of these genetically modified organisms (GMOs).

While the introduction of transgenic soya reduced costs, it generated a dependence on behalf of farmers that previously seldom existed, subordinating them to a very concentrated market, dominated by five large firms on the world level.

Social Consequences of the Main Transformations of the Last Decade

The processes and transformations we describe above had an important social impact on the AFS as a whole: urban and rural employment was substantially reduced, and the situation for small and medium independent farmers, and access to food for vast sectors of society, deteriorated substantially.

Destruction of Jobs

Transformations in the agro-food system contributed to the expelling of large numbers of workers from this sector. While in the 1990s agro-food industrial production and retailing increased, employment fell. The gap between production and employment was characteristic not only of the agro-food system, but also of the economy as a whole.

In the 1990s, and in particular after 1995, employment in the food industry and in final retailing of food fell significantly. Comparisons of the Census of 1985 with that of 1994 show a strong reduction of employment in the 'food, beverages and tobacco' branch of industry from 360,630 persons employed (workers and employees) in 1984 to 275,661 employed in 1993, implying a 24 per cent reduction (84,969 persons) (INDEC, 1995, 1987). Recent INDEC data projections show that by the year 2000 only 225,700 persons are expected to be employed by the 'food, beverages and tobacco' branch of industry. This means that in a six-year period (1993-2000) the food industry contracted by some 50,000 employees.

There are many causes for this reduction in employment. Increased concentration and vertical integration imply new organizational forms that usually tend to reduce employment. Technical advances in industry, and changes in the industrial structure with a larger progress of branches that are less employment-creating, is another factor. Among those branches of industry that in comparative terms increased production substantially is the oilseeds complex, one of the main agro-industrial complexes of Argentina. While this agro-food complex exports about 40 per cent of Argentina's total agriculture and food exports it employs only 5,600 workers in the industrial stage, a very small proportion indeed of employment provided by the overall food industry. It is evident that despite the growth of this industry in recent decades it is not a significant generator of employment.

As is pointed out above, apart from reductions of employment in the food industry, there has also been a substantial decrease in employment in retail distribution. Between the Censuses of 1984 and 1993 more than 64,000 small retail distribution shops disappeared implying a reduction of 125,000 employed persons. These reductions were only partially compensated by 22,000 new work positions created by the supermarkets (SAGPyA, 1997). These tendencies continued in more recent years. On the whole in the 1984-1998 period total employment lost in food retailing affected 255,000 persons; in this period 150,000 work positions of wholesalers were also lost. (SAGPyA, 1997). Thus, the intense 'supermaketization' of wholesale and retail food distribution in Argentina was an aggressive destructor of employment.

How Small and Medium Farmers Were Affected

Farmers and farm labour were also affected by these transformations. Deregulation and changes that took place in the agro-food system reduced the decision-making capacity of farmers, the inputs they use in production creating new forms of dependency, and their negotiating capacity over prices and other conditions affecting their sales. Deregulation also created greater instability. Farmers now have to cope with large variations and instability of international food prices which directly influence domestic prices.

Increased concentration meant that industry and/or supermarkets, the 'integrative loci' in each complex, were able to reduce agricultural margins, impose quality conditions, as well as forms of presentation and of transportation of the commodity. And – as was shown above in the case of the potato complex – the supermarkets even managed to impose the variety and inputs to be used by the farmer.

Due, in part, to these reasons and to the strong pressures from industry and supermarkets, many small independent farmers disappeared – obliged to emigrate to the city or to change their occupation.

CONINAGRO (Confederación Intercooperativa Agropecuaria, Cooperativa Limitada), the organization that represents farm cooperatives, estimated that in the Pampaen Region some 30 per cent of small and medium farmers stopped producing between 1992 and 1997. A similar trend (26 per cent) was estimated for independent sugar cane farmers of the Province of Tucumán, a traditional regional crop (Giarracca *et al.*, 1998).

Thus, we are confronted with several causes for the reduction of the small and medium independent farmers and peasants in Argentine agriculture. Liberalization policies, and deregulation measures contributed to wide commodity price variations and instability. Increased negotiating capacity of industry and supermarkets (due to the concentration and vertical integration processes) vis-à-vis the farm sector, influenced prices and conditions of production, and purchasing conditions for inputs. An increased technological dependence on behalf of the farmers with regards to suppliers of seeds and fertilizers also contributed to these trends. An important factor that should be highlighted is access to credit; the elimination of special credit facilities, in many cases previously facilitated by the state, imposed greater financial costs on the small farmers. All these factors have negative impacts on small and medium farmer and in many cases contributed to their disappearance as such in agriculture.

Conditions Affecting Access to Food

Argentina is one of the few surplus-food producers of the Third World, and remains one of the main grain exporters to the world economy. Consequently, it does not sustain the external food dependency that characterizes most Third World countries. Despite this fact, hunger and poverty have increased in recent years denoting the increased difficulties for maintaining food entitlements for vast segments of the population (for the concept of entitlements see Sen, 1981).

Historically, Argentina had a relatively cheap food economy in which access to food for vast social sectors was widespread. In recent decades food prices (in real terms) have tended to increase on the domestic front. It is interesting to notice that these price increases, at least in the 1990s, were to some extent due to the increase in distribution margins – that is, the difference between wholesale and retail prices – which denotes one of the impacts of increased supermaketization. These trends occurred even in periods of falling agricultural prices (Teubal *et al.*, 1995; Teubal and Pastore, 1998; Teubal, 1999).

Increases in relative food prices can be partially ascribed to the greater concentration and centralization of capital, and vertical integration, and to other transformations to which we have made reference above. Needless to say, these increases in food prices had a bearing on the reduction of food entitlements, in particular for the lower income segments of society. Other factors reducing food entitlements of these sectors of society were increased unemployment, and lower income and wage levels. Both were a consequence of the structural adjustments applied drastically in recent decades, but in particular in the 1990s.

References

Arcor (2000), Official Website, www.arcor.com.ar.

Barraclough, S. (1982), *A Preliminary Analysis of the Nicaraguan Food System,* UNRISD, Geneva.

Fernández, A. M. (1997), 'Reestructuración del Subsistema Lácteo Argentino: dos Décadas de Modernización Concentradora', *Realidad Económica,* (Buenos Aires) vol. 146, pp. 104-119.

Friedmann, H. (1992), 'Distance and Durability: Shaky Foundations of the World Food Economy', *Third World Quarterly,* vol. 13(2), pp. 371-383.

Giarracca, N., Gras, C., Gutiérrez, P. and Bidaseca, K. (1998*),* *Conflicto y Negociación en el Contrato Agroindustrial: la Actividad Cañera Desregulada en Tucumán,* Informe de Investigación, GER, Universidad de Buenos Aires, Buenos Aires.

Heffernan, W. (1998), 'Agriculture and Monopoly Capital', *Monthly Review,* vol. 50(3), pp. 46-59.

INDEC (Instituto Nacional de Estadísticas y Censos)(National Institute of Statistics and Census)(1987), *Censo Nacional Económico 1985. Industria. Versión definitiva,* INDEC, Buenos Aires.

INDEC (1992) *Censo Nacional Agropecuario 1988. Versión definitiva,* INDEC, Buenos Aires.

INDEC (1995), *Censo Nacional Económico 1994, Industria. Versión definitiva,* INDEC, Buenos Aires.

INDEC (1999a), *Grandes Empresas en la Argentina 1993-1997,* INDEC, Buenos Aires.

INDEC (1999b), *Comercio Exterior Argentino. Resumen Estadístico 1997-1998,* INDEC, Buenos Aires.

INDEC (1999c), *Industria Manufacturera. Producción, Ocupación y Salarios 1990-1999,* INDEC, Buenos Aires.

Mateos, M. y Capezio, S. (1999), *La Consolidación de las Cadenas Multinacionales de 'Fast Foods' y el Impacto de la Producción bajo Contrato en los Productores Primarios,* Paper presented to Primeras Jornadas Interdiciplinarias sobre Estudios Agrarios y Agroindustriales, 4 and 5 November, Buenos Aires.

SAGPyA (Secretaría de Agricultura, Ganadería, Pesca y Alimentación)(Secretary of Agriculture, Livestock, Fishing and Food)(1993), *Estudio sobre Competitividad,* SAGPyA, Buenos Aires.

SAGPyA (1996), *La Industria Argentina de Alimentos y Bebidas,* SAGPyA, Buenos Aires.

SAGPyA (1997), *Informe Sector Azucarero,* SAGPyA, Buenos Aires.

SAGPyA (2000), *Estimaciones Agrícolas – Cultivos Industriales,* www.sagpya.mecon.gov.ar.

Sen, A. (1981), *Poverty and Famines,* Clarendon Press, Oxford.

Teubal, M. (1998), 'Transformaciones en el Sistema Agroalimentario: Impacto sobre los Precios Relativos', *Informe de Coyuntura* (Centro de Estudios Bonaerense, La Plata) Año 8, vol. 77, November-December, pp. 77-87.

Teubal, M. (1999), *Complejos y Sistemas Agroalimentarios: Aspectos Teórico.Metodológicos,* in N. Giarracca, (ed.), *Estudios Rurales, Teorías, Problemas y Estrategias Metodológicas,* La Colmena, Buenos Aires, 97/121.

Teubal, M. (2000/2001), 'Structural Adjustments and Social Disarticulation: The Case of Argentina', *Science & Society,* vol. 64(4), Winter, pp. 460-488.

Teubal, M. and Pastore, R. (1995), *El Agro y los Complejos Agroindustriales: El Caso Argentino,* in M. Teubal *et al.,* (eds), *Globalización y Expansión Agroindustrial.¿Superación de la Pobreza en América Latina?* El Corregidor, Buenos Aires.

Teubal, M. and Pastore, R. (1998), *Acceso a la Alimentación y Regímenes de Acumulación. El Papel de los Precios Relativos,* in M. Teubal, (ed.), *Teoría, Estructura y Procesos Económicos, Ensayos en Honor al Dr. Julio H. G. Olivera,* Buenos Aires, Eudeba.

Teubal, M. *et al.,* (1995), *Globalización y Expansión Agroindustrial.¿Superación de la Pobreza en América Latina?* El Corregidor, Buenos Aires.

Vigorito, R. (1979), 'Criterios Metodológicos Para el Estudio de Complejos Agroindustriales', in *El Desarrollo Agroindustrial y la Economía Internacional,* Documentos de Trabajo para el Desarrollo Agroindustrial, vol. 1, Coordinación General de Desarrollo Agoindustrial, SARH, México D.F.

Chapter 9

Technological Transformation of Household Milk Production in the South of Brazil: Social Impacts

Ivaldo Gehlen

Introduction

This chapter is the result of research into the impacts of technological transformations and market determinations on the socio-professional identity of household producers of milk in the State of Rio Grande do Sul. These transformations alter both social relations and the very concept of such terms as 'labour'. They also introduce new ones, for example, in the areas of quality and production, and affect socio-professional identity, especially among those who either still hold to their traditional system of production or find themselves in transition to modern production.

The study is grounded in a review of the literature, in semi-structured interviews and in questionnaires conducted in five milk-belts in Rio Grande do Sul. The data were used to analyse the conflict between two types of competitive rationality – that of the agroindustries and that of the producers. Agro-industries hold productivity as their priority while producers, organized in scores of associations and cooperatives, prioritise social reproduction together with the agricultural workers' social, cultural, economical and environmental sustainability in an effort to discourage the process of their exclusion and to promote a better quality of life. While this debate is located in Brazil, it is argued that its theoretical content may be applied elsewhere.

The southern region of Brazil, constituted by three States, is the second major milk producer within its five geopolitical regions of the nation. It is responsible for the production of nearly 3.2 billion litres a year. According to the 1995/96 Agrofarming Census the total milk production in Brazil reached nearly 19 billion litres a year. The State of Rio Grande do Sul is the major milk producer in the region and the third within the 26 Federation States. Production amounts to nearly 1.4 billion litres a year, with an average production of less than 50 litres a day per producer. According to figures by FAO (DESER, 2001, p.8) some 93 per cent of those producing milk are household producers.

The consolidation of all the segments of the Agroalimentary Chain of Milk in the South of Brazil has been so dramatic that it approaches oligopolization, which triggers a selection process among the producers. The social impacts of this process affect labour organization, technological practices, and relations with the market and with consumers, not to mention the survival of over 200,000 milk producers in Rio Grande do Sul. The great majority of these household milk producers are involved in more than two commercialized agricultural products (many of these households are also pluriactive, where family members perform activities other than the agricultural ones).

Partly as a reaction to the social impacts of the technological transformations, producers are adapting concepts such as 'competition' in seemingly innovative ways. As a result, competition is analysed using the dimensions of quality and productivity. These two dimensions – besides affecting the household milk producers in the region – have a strong interface with the policies for the sector. An example of this is the new legislation on sanitation. As well, the interface between producers and the institutions geared to the sector must be considered. This eventually determines the reorganization of family labour and influences decisions about the production system – technological alternatives included. This set of situations obviously affects socio-cultural identity – primarily professional identity. The majority of producers hold to the traditional system of production and diversify their production to include cereals and small animals, while combining production for the market and for domestic needs. With few exceptions, milk is characterized as a non-strategic product; it is considered only complementary or even supplementary to other production. Therefore, the adoption of milk as a strategic or major product leads to the necessary (re) professionalization as a condition for permanence. This change in the professional identity also affects, to a certain extent, the original socio-cultural identity.

In order to understand this process it is necessary to analyse the ideal of competition and its multifunctional roles in the present context. The formulation of such a process is rooted in history and is a result of the experience of different social actors, mainly those in the industrial sector, and of the analysis of several authors who seek to explain and/or characterize the dynamics of labour today. Likewise, there is an imposition of this ideal on other social actors – such as in the case of household milk producers. Thus, the analysis of the impacts provoked by the very ideal of competition is momentous. The multiple approaches to this ideal are shaped by the social actors' specific agendas, which correspond to different rationalities of production systems within, and labour relations among, household producers. The coercive imposition of the modern producers' conception of productivity by the agro-industries, cooperatives, associations and public policies (as it is the case, for example, of the new Brazilian legislation on milk quality) is very powerful.

The debate on competition in Rio Grande do Sul is not complete without considering the relationship between the milk chain agricultural workers and both the state and private institutions. The problem concerns the producers' practices on the one hand and the production and diffusion of the ideal of competition by the institutions on the other. This very ideal of competition generates social differentiation and re-stratification among the agricultural workers, allowing some to be strengthened and others to be excluded – such are the impacts of these relations.

The Issue of Competition

The concept of 'competition' – as something desirable, and to be pursued – accords with the economists' and marketers' view of the world. Its practical manifestation is in the economic transformations of production. From the beginning of the 1970s a 'new model' of development was promulgated throughout Brazil. In this model, the State was gradually divested of its developmentalist and/or interventionist character. In Brazil, these changes have been associated with the pressure exerted by international organisations (such as IMF and the World Bank) together with the growing internationalization of enterprises – the latter occurring as a result of the purchase of successful (competitive) Brazilian enterprises by multinational companies.

The implementation of policies that endeavour to consolidate this new model cannot be dissociated from external pressures from multinational companies and financial organizations. Implementation of such policies took place in the beginning of the 1980s, coinciding with the so-called crisis of Brazilian foreign debt – increasing the nation's financial dependence on external funds. Today, the strongest pressure is to enforce the transfer of the role of intervening directly in the production sectors from the State to private enterprises, along with the State's eventual retreat from mediating the labour-capital relationship. A series of deregulation policies have been put into practice; for instance, that of the price of milk to the end consumer (which had formerly been controlled), the policy for equipment acquisition, and the importation policy. It is important to bear in mind that, in the Brazilian case, as well as the other Latin American countries, the state exerted – and still does to a considerable extent – an interventionist policy in economic development. The recent changes are based on only a superficial state reform as both a formulator and executor of global and sectorial development policies. The state is gradually implementing the processes fostering globalization, and endorsed by the G7 countries (US, Japan, France, England, Germany, Italy, and Canada).

Competition, generally speaking, is posited as a goal that guides the process of production restructuring. The opening of the economy to the international market enforces the adoption of productivity and quality standards determined by such market criteria. A negative reaction on the part of domestic enterprises (failing to meet the required demands) jeopardizes their survival as enterprises. Having incorporated features other than those regarding production efficiency and the price of products, the ideal of competition itself is in the centre of production redefinition. Product quality, environment and the 'quality' of the very conditions of the production process, and the qualifications of workers are now part of this concept. Competition is being measured according to performance levels, efficiency, quality, and the training of workers for operational ends.

In this sense, competition is increasingly related to the productive strategies, in order to meet the new standards exacted by the consumer/market. Productivity, as long as it is part of competition, is not the result of the efficiency of labour only, but also incorporates capital. Therefore, the notions of efficiency and of flexibility have to do with satisfying the different and differentiated demands, that is, innovation.

In Brazil, increasing competition within the production chains represents a strategic challenge to development and to the affirmation of its presence in the international consumer market. Brazilian participation has traditionally been due to low value-added – therefore cheap – products, both industrial and agricultural, which were produced with little technology and with the use of a barely qualified and low-wage labour force. Adding to this, the domestic market's (the Brazilian consumer's) self-image persists as a rather sloppy one, resulting in no need for further investment in quality. In the 1990s this image began to undergo a process of reversal through the importation of some mass-consumption products as well as via legislation. The latter encompassed the consumer defence legislation and the consumer's 'qualification' by means of wider access to information and education. In the case of milk, this change was evidenced through the improvement and diversification of dairy products, the incorporation on new technologies, and the change in the legislation on quality – while it determined procedures. This was especially evident in agriculture production and in the transformation segments – in the agricultural chain.

Resuming the original focus on the sectorial public policies, the relevant role they have performed must be acknowledged. This role encompassed stimulating activities or guaranteeing their continuity and establishing classification standards for products. The criteria for these standards are often the result of pressures on the part of segments of the chain – particularly the transformation segment, which has been increasingly controlled by multinational companies. Italian Parmalat alone purchased 14 Brazilian milk industrialization enterprises between 1990 and 1998 (DESER, 2001) making it the biggest company of its kind in Brazil. Rio Grande do Sul has the lowest rate of internationalization in the transformation segment, when compared to the other Federation States. Nevertheless, the process of oligopolization has led to the control of about 75 per cent of milk collection and industrialization by two enterprises only: the Brazilian Elegê Alimentos and the Italian Parmalat, the former holding some 50 per cent of the market.

The growing dominance of the agro-industries over the farm sector also generates new social conflicts and new forms of differentiation and social inequalities. The entrepreneurial strategy of presenting itself as bearing a 'typical identity' – which is neither that of an industry in the classical sense, nor that of an agro-industry – endows it with a certain ambiguity. This ambiguous stance enables it to manipulate the situation according to specific interests. In so doing, the enterprise escapes social visibility and control, and keeps the forms and relations of domination hidden, both above the segment (bottom-up) – with the raw material producers especially – and below it (top-down) – with the consumers.

Bottom-up, the enterprises transfer the conflicts with the producers (on price, technology, quality criteria, and so forth) to the domain of the producer-technician. Technicians are the executors of the agro-industrial or developmental policies, along with the institutions that generate and diffuse technology and knowledge. This is why some social struggles on the part of producers are directed against the equipment manufacturers and the banks (The Bank of Brazil, in this case). On several occasions the agricultural workers have stood on picket lines before the bank agencies and prevented them from opening. Such procedures have proved of little benefit but have grown to have a strong symbolic effect on public opinion and on governmental agencies. Recently, the milk producers in the South of Brazil have adopted sit-ins at the agro-industries' premises, thus gaining visibility in the media. In doing so they have sought to earn legitimacy from consumers.

Top-down, agro-industries are articulated with the dealers and the consumers, managing to blame the high prices of dairy products on production cost and taxes. They have also been able to impose their concept of quality. In an attempt to reduce the producers' bargaining power, the exigency of adopting new technologies, genetics, and even rational procedures is passed on to them by the agro-industries, disguised as demands from consumers. The 'all-powerful consumer' is presented to the milk producers as having a generic and abstract identity, in many cases a substitute for the concept of market. In such a top-down relationship, the *dealer* plays an important role to press the agro-industries for quality and price. The dealer is the consumer's 'representative', in Pierre Bertrand's words (1995). According to Bertrand, the consumer today questions the social and environmental impacts of consumption society, when it is the very presence of the consumption society that should be questioned.

As the producers' unions (mainly the household producers) have overlooked the issue of the relationship with the consumer, it has successfully generated the desired effects anticipated by the agro-industrial and distribution enterprises. They have successfully manipulated ideologically the consumers' power in the consumer-producer relationship. The attempt of a reversal of this effect through the establishment of a producer-consumer alliance for the sake of citizenry and quality of life is still a very recent one, which makes it difficult to be properly appraised.

There are many factors and agents defining the quality and productivity standards as 'competitive'. Among the requirements internal to the agro-industrial milk chain are technologies such as automatic milking, coolers and facilities, concentrated meals, veterinary and agronomic assistance, transport and remedies. The necessity for these technologies is gradually imposed onto the producers so that eventually they believe the ideological notion that only some of the producers fulfil the competitive conditions necessary to succeed. The rest are faced with the ever-present menace of being excluded from the system.

This threat to the household milk producers in the South of Brazil finds its antidote in their growing organization into cooperatives, unions, and associations. These organizations have engaged in social struggles in an ever more articulated way. Pressed by the producers, dissent is expressed by a variety of means: documents, media and mass demonstrations, such as parades on motorways, and sit-ins at industries and bank gates. They express their demands and/or propositions concerning prices, supportive public policies, local infrastructure such as roads and schools, and the refusal to accept harmful standards.

Global competition and 'the emergency of a coalition of actors committed to redefine properties for the regions, based on the productive potential of the resources within the household production system' (Wilkinson, 1996, p.177) can provide *niches* for special products like cheese, organic milk, and yoghurt. This alternative, however, is a restricted one in the Brazilian situation due to two factors: first, the low purchasing power of the population; and second, the tradition that undervalues these products. This leads to the forecast that such alternatives will find a quick saturation. Nevertheless, the so-called 'informal' practices typical of small co-operatives or household agro-industrial units, have been growing (the case of non-fiscalized milk in Brazil). This can still be expanded despite the restrictive character of the recent legislation (known as 'milk quality' law). The option to fill these *niches* is possible because total milk production falls short of satisfying the existing domestic demand – about 10 per cent of milk consumed is imported. The technological alternatives available facilitate agricultural production and its industrialization. These alternative practices may grow to the benefit of the household producers, as long as formal or informal arrangements between producers and consumers are mutually beneficial.

Characterization of the Milk Sector in Brazil and in Rio Grande do Sul

Brazilian milk production nearly doubled in the 20 year period from 1980 from a little more than 11 million, to over 20 million, litres a day. It is worthy of notice that the production of the so-called informal milk (so called 'non-fiscalized') exhibited the highest increment, from about 3.5 million to 8.5 million litres a day in the period mentioned above, whereas formal (fiscalized) milk increased from 7.7 million to 11.5 million litres a day (IBGE, 1990-97).

The growth in milk production is continuous, ranging from between 4.0 and 4.5 per cent a year. These figures mirror the overall increase of milk and dairy products consumed by the Brazilian population. Compared to other countries, Brazilian milk production qualifies as a low-productivity one. Such low productivity is the result of the history of the introduction of milk production in Brazil. As beef cattle were assigned priority, milk production has rested on the traditional system, which is still predominant among its producers (not only the household ones). Adding to this, the urbanization process is a relatively recent one (starting in the 1950s) and the process of change has proved a slow one. The permanence of the traditional producers is at the same time a problem and a potential. The potential exists in the preservation of a broad social basis that may ensure both expansion and sustainability to the sector.

When comparing the regions in Brazil, the profiles of the producers as well as productivity differ. The producers in the Southwest and Mid-West regions have higher production rates, as they are predominantly middle and large-size. As a consequence, the 5.4 per cent that produce more than 200 litres a day in the Southwest are responsible for 41.3 per cent of total production, whereas 73.1 per cent of those with smaller production contribute only 21.1 per cent of the total in the region. In the Mid-West the distribution percentage is similar (IBGE–Agrofarming Census, 1997; Associação Gaúcha de Laticínios [AGL]).

The characteristic shared by all the regions is the major presence of producers with low daily production, that is, less than 50 litres a day. In the Northeast, the North, and in the South over 90 per cent of the producers constitute this strata, and contribute more than 50 per cent of the production in each region. Rio Grande do Sul bears a profile similar to the regional one. However, the formal, fiscalized, milk production tends to be concentrated in a smaller number of producers. These are the ones selected according to efficiency, as the traditional producers with low production are excluded. In Rio Grande do Sul those who produce more than 100 litre a day represent 11.38 per cent and contribute 42.47 per cent of the volume yielded. On most properties in the State, milk is still a secondary activity generally handled by women with the help of their children.

The Changes in Household Milk Production in Rio Grande do Sul

The concept of a production chain is analytically appropriate in order to understand the processes, the dynamics, and the relation between the social agents involved in the production, industrialization and commercialization of milk in the State. The concept of chain only makes sense when considered within the agro-industrial complex, which shows the multiplicity of relations involved in agricultural activity. These include state policies, diverse agents' organizations, research support, besides the so-called production chain itself. The idea and the concept of the agro-industrial complex can only be operationalized when some definitions are elaborated with methodological accuracy. Among these, are the way in which the chain is organized for each product, the activity of the social actors whose interests focus on such a product, the role of the State, and consumption. For Graziano da Silva (1991), the agro-industrial complex is an articulation of interests originated both in productive activity and public policies. According to this concept, the actors and agents are divided into chain-rings according to their attributions and roles regarding the product. These agents are mutually interconnected to constitute what is known as a chain. The concept of chain makes it possible to understand the heterogeneity that is typical of the rings, defined by situations of convergence and of conflict alike.

Therefore, the analytical relevance of the notion of chain for the studies on milk production lies in the fact that it embraces the understanding of the political, economic and social discrepancies inherent to the clash of interests of the agents within the production process. In the case of milk production these differences are often expressed in the form of conflict, especially in the relations between the agro-farming and transformation rings.

In such a scenario, marked by an heterogeneous and differentiated distribution of the conditions to the agents in the chain rings, the incentive measures and policies informed by the neo-liberal logic tend to be beneficial to the large agro-industries rather than to the agricultural workers. This eventually leads to the exclusion of those deemed to be less competitive, and transfers the negative social impacts of this process to society in general.

The small-sized producers have associated the imposition of new techniques and technologies with the aggravation of their situation. The large agro-industries' endeavour to maximize production by determining standards that preserve product quality has proved harmful to these producers. The high cost of production associated with their low-income milk-business will force many of them to switch to other economic activities.

Among the factors that have significantly contributed to the changes in the structure of the milk production chain have also affected the State. An agent exogenous to the production chain, while still maintaining a strong power of determination over it, the State has turned into a passive agent. It allows the free course of interests associated to the market logic without offering any controlling or countervailing policies. In this way, the State fails to play its function as social regulator; nonetheless, it will be pressed to take on the responsibility for the search for alternatives to counter the negative social impacts generated, while not blaming those responsible for their generation.

The Household Producer's Identity: The Case of the Milk Producer

In the context of the significant changes in the milk agro-industrial complex in Brazil, the milk production chain in Rio Grande do Sul is hit by reflections that affect the socio-cultural identity of the families of agricultural workers involved in production. The introduction of new practices and technological formats in production has had the same effect, leading to a process of social mobility. This is expressed in either their inclusion or exclusion in the agro-industrial chain and therefore their 'place' in local and/or global society.

It is interesting to investigate the impacts on the household milk producers' identity as a result of their introduction into a new production chain whose dominant rings try to be competitive. Social identity is taken here as one of the social forms to gain recognition for what one is, thereby affirming social dignity.

The milk producers that were interviewed are all of the household kind and the majority of them own their land. The average property size is approximately 20 ha, ranging from 5 ha to 58 ha. The number of cows per property varies considerably. Where milk is a priority, cow density is larger. The smallest property (5 ha) had 20 cows and the property with the most cows (50) was 47 ha.

As far as the use of technology is concerned, there is a broad diversity. In the milk-belt of Teutônia – a modern and consolidated one – the standard is high: 96 per cent make use of artificial insemination and 93 per cent have automatic milking. In the milk-belt of Três de Maio, the traditional system prevails.

The increase in productivity depends, thus, on the place of milk in the productive unit and on the professional activity of the manager. In the milk-belt of Teutônia 20 per cent of the interviewees wish to make milk production their exclusive activity. A further 53.3 per cent would like it as their main activity; more than half of them have already taken courses to qualify for the activity. In the rest of the State, milk production prevails as a complementary activity and less than 20 per cent of the interviewees have been trained or joined a qualifying program for the activity.

The definition of milk quality – conditions of production included – is broadly stated as reduced number of micro-organisms, low acidity after pasteurization or sterilization, resistance to temperature, high protein and total solids content, and absence of antibiotics. This definition serves as the standard for the selection of producers by the agro-industries. The selective pressure on the producers is made through the payment of differentiated values according to quality indexes, to the quantity of litres per producer, and to the regularity in production. There are still some differences in freight payment depending on the quantity delivered. As the quantity delivered also serves to define the frequency of collection, not only the purchase of coolers is made necessary, but professional training is also required.

The imposition of these and other (economic and technological) objective conditions in the production system cannot be counted as the only factors to generate the threatening exclusion of household producers from their activity and socio-cultural *milieu*. There are also subjective conditions that seek to intervene in their system of values, images, symbols, and even in their private social relations (especially by 'bureaucratising' relations through the creation of a countless numbers of associations). Such changes are guided by a certain dialogue and by examples that exalt the competitive efficiency of the modern producer together with market rationality. The market is portrayed as an omnipresent and still unattainable being, immune to pressures on the part of individuals or local organizations.

This mechanism de-structures the traditional strategies and removes the initiatives of resistance on the part of the household agricultural workers. The milk producers of Rio Grande do Sul – about 80 per cent of which are co-operative members – have accumulated an important experience in the social struggle for their interests. Firstly, they have been organized in co-operatives (created in the early Twentieth Century) – secondly, and starting in the 1960s, they have joined rural workers' unions. More recently, they have created and supported hundreds of associations of various kinds.

The majority of the agricultural workers interviewed for this study believe that it is possible to interfere in the sector policies and expressed a willingness to engage in social struggle. Such an engagement is geared to change the product price, to intervene in the legislation deemed as discriminatory (and openly favourable to the agendas of big business and producers), and also to create the means capable of adding value to their work. As an example of this, the team purchase of equipment and even the industrialization in co-operatives are mentioned. This socio-cultural patrimony of household agriculture engenders new forms of organization of milk production. It articulates the competitive rationality imposed by the hegemonic agents in the industrial segment today with its own tradition of identity and solidarity. It even incorporates new elements within the concept of competition, such as social reproducibility and environmental sustainability.

Such impacts are considered in a context that prevents the household agricultural worker's identity from being reproduced. In this dynamic, the household agricultural worker's identity, history and the socialization that (re)produces identity vie for centrality with other social positions (Champagne, 1986). In this sense, being a household agricultural worker may be one project among others, especially within the younger generation. In order to understand such a process it is necessary to examine the influence played by the workers' introduction into the milk production chain, with its quality and productivity standards, on the household agricultural workers' social practices. In addition to the framework that has been analysed here, there are others that have to be considered so that other possible social positions can be envisioned. Consider the influence played by the State, by education (which brings in other knowledges and worlds), by the migration of family members, by work outside agriculture (pluriactivity), among others. The availability of communication and contacts with other social *milieus* generates comparisons between the household agricultural worker and other social positions. All this influences decision-making, for the permanence (or not) within the category of a household agricultural worker, and for the definition of the priorities for the production system.

The socio-cultural and professional identity of the household milk producers in the South of Brazil finds itself in a permanent dynamic relation with economic, socio-cultural and environmental factors. The producer lives in permanent 'tension' concerning the perspectives of social and economic reproduction. Such dynamics interfere in his/her standpoint before the exigencies made by the agro-industries (competition for productivity and quality) and in relation to the producer's collective identity as a household agricultural worker. In such an internal relation of identity, collective action becomes important, as it is through this action that its members can understand one another.

In this case, an outstanding change in their professional identity becomes evident, since efficiency imposes procedures formally learned and a specific rationality – one almost exclusive to milk producers – which leads them to abandon generalism and traditionalism. Regarding socio-cultural identity, the transformations in progress affect these producers in different ways. For most of them, adjustments are made in order to ensure the continuation of the family household. The traditional and the innovative forms of organization and the re-valuation of customs and traditions strongly point to this evidence. For a minority, the transformations in progress lead to a process of vertical social mobility, that is, one of change in their class condition. This minority seeks inclusion into the agro-industrial chain; their agenda lies in a rationality that leads away from the traditional one. The ideal type of reference seems to be the capitalist entrepreneur model, but there is not enough data available to reach a reliable conclusion.

What can be recognized is that some of the household milk producers are being excluded in this process. Exclusion is coerced through the imposition of techniques and technological conditions relating to production, and through the imposition of new values, symbols and rational habits that alter the traditional ones. This process of exclusion 'works' by exalting competition (as conceived of by the ruling market rationality), by the contingent adoption of a 'new' professional identity, adjustment of socio-cultural identity, and by changing the representations of the 'lived' reality of producers.

When industrial rationality is predominant, the visible impacts are manifested in a plethora of measures. The industrial rationality centres upon adding value to products according to the determinants of the transformation and distribution rings of the milk chain. The visible face of such predominance takes several shapes. Free pricing of milk and admission of new agro-industrial companies into the sector is one of them; compliance with a new set of market rules, another. Minimum production (per producer and per cow), implementation of costly new equipment (cooler, automatic milking), professionalization of labour, and strict obedience to sanitary rules for flock handling (subject to law enforcement) are all unfolding aspects of industrial rationality.

The strategy to fight exclusion from the milk chain lies on both the incorporation of low-cost technological resources and on associative and community-based actions. These actions aim to optimize gains by selling and buying in groups, by pressing for fair prices for products, or by creating cooperative systems for milk collection, transportation and industrialization in some cases.

The producers' professionalization, the strategic centrality of milk on the properties, the adoption of innovative technology for production, genetics and handling of the flock are *sine qua non* conditions to warrant survival in this activity and non-exclusion from the production chain. This leads to the conclusion that the process of transformation is not reversible. The guiding *conception* of this process is reversible, though. Here lies the essence of this clash. On the one hand, there is competition as purely economic efficiency (productivity aiming at profit and accumulation) leading to the regulation of production, sanitation, and transportation methods among others. On the other hand, there is competition as socio-economic efficiency and attuned to social and environment issues. The latter takes quality of life as priority, leading to the adoption of a cooperative rationality, a focus on the economy of the property's internal resources, and new techniques that aim at social sustainability and reproducibility rather than at productivity.

Countering the concept of productivity is that of sustainability. The latter can be employed both as a highly effective strategy to develop resistance to the sector as well as to provide a safe model that fits the household producers' tradition. Instead of putting their tradition at risk, such a model should satisfy consumers in terms of quality and quantity, even in face of the growing demands in the milk sector.

Research in the State of Rio Grande do Sul has produced some initiatives to re-conceptualize competition in the light of sustainability in a broad sense. Such initiatives were raised in producers' co-operatives or associations to confront the big agro-industries. Some of these initiatives are as follows:

- Assign milk production strategic centrality, restructuring the production system accordingly and re-distribute household labour-force responsibilities, that is, assign this activity, which has traditionally been handled by children as an extension of their domestic activities, to the major labour force (often male). This option corresponds to the rational reorganization of the production system, integrates specific organizations, and acknowledges professional training;
- Associate milk production with other commercial cultures, preferably those that do not imply a dispute in the deployment of land and labour. The option for other products depends on the region, on the structural

conditions and on the family members' own wishes, eg maintain production to satisfy domestic needs;

- Improve progressive productivity gains, establishing flexible productivity floors and valorize to the maximum the resources in place in the property, such as genetics. In the case of feeding, prioritise grazing (animal collects the meal) and implement rotating systems to diminish the use of supplied meals – 40 per cent maximum. The conventional modern industrial system recommends about 60 per cent of meals be supplied to the cow (hay, forage, concentrated meals and so forth);
- Provide alternatives to existing co-operation (associations) to take in the new technologies and changes in the production system, such as group purchase of equipment, exchange of genetics and varieties of food, and sharing resources, utilising associative forms or groups to maximize technical assistance, and sharing successful experiences; and
- Develop training programs for milk producers using a variety of means (courses, lectures, field days, technical schools, audio-visual pieces, radio, etc). This education program would be coherent with the notion of chain and entertain ideas of market, agro-ecology, sector policies, and competition. It would promote the objective of professionalization of milk production, and of the entire production chain.

Conclusion

The modernization of Brazilian agriculture generates contradictions, including its current economic, social and ecological unsustainability. The adoption of alternative technological formats makes the social reproduction of the household agricultural workers possible, and enhances their inventive capacity. Organic production, increasingly accessible to these producers, tends to subvert the logic imposed upon the household producers by the big agro-industries: it constitutes a site of resistance to this logic. The organic product's distinctiveness lies in both the commitment to the preservation of the socio-cultural identity of the rural producer, and on its quality as healthful and beneficial to the consumer.

The producers who were analysed in the scope of this study are still perplexed at the transformations in progress, but they are politically organized and looking forward to sharing productive solutions: community coolers, rotating credit, bartering, technical assistance in groups, technology exchange, and micro-regional milk-belt planning. They systematically put forward public actions and struggles that demand mobilization and engagement. The result of these efforts is more dependent on the degree of the producers' mobilization and engagement than on exogenous technological factors.

Acknowledgements

The author wishes to thank Gabriele dos Anjos, Sayonara Araújo, Vilso J. Antonello, as well as Alexandre S. Medeiros and Maurício Russo for their collaboration. The author also acknowledges the support of FAPERGS. The translation into English was undertaken by Isabella Vieira de Bem with further assistance from Janet Norton.

References

Aguiar, R. (1986), *Abrindo o Pacote Tecnológico*, Polis/CNPq, São Paulo.
Basso, D. and Silva Neto, B. (1999), 'Controvérsias sobre Profissionalização e Desenvolvimento na Agricultura: O Caso da Produção de Leite no Rio Grande do Sul', *Indicadores Econômicos,* vol. 6(4), pp. 232-246.
Batalha, M. (1997). 'Sistemas Agroindustriais: Definições e Correntes Metodológicas, in M. Batalha, (Org) *Gestão Agroindustrial,* vol. 1, Atlas, São Paulo, pp. 23-48.
Bertrand, P. (1995), personal communication.
Canclini, N. (1997). 'Consumidores e Cidadãos: Conflitos Multiculturais da Globalização', UFRJ, Rio de Janeiro.
Carciofi, R. (1997), 'El Papel del Sector Público en Desarrollo Latinoamericano', *Revista de la CEPAL,* vol. 6, pp. 7-15.
CEPEA and FEALQ (1998), *Boletim do Leite* 5, p. 48.
Champagne, P. (1986), 'La Reproduction de l'Identité', *Actes de la Recherche en Sciences Sociales,* vol. 65, pp. 41-64.
DESER. (2001), *Boletim do Deser,* vol. 117, pp. 5-14.
Fernandez, D. (1995), 'Programa do Setor Leiteiro do RS no Âmbito do Mercosul', EMATER/RS, Porto Alegre.
Fernandez, D. (1995), *Diagnóstico do Setor Leiteiro do Rio Grande do Sul no Âmbito do MERCOSUL*: EMATER/RS, Porto Alegre.
Franco, R. (1996), 'Los Paradigmas de la Política Social en América Latina', *Revista de la CEPAL,* vol. 58, pp. 9-22.
Gehlen, I. *et al.,* (2000) *Os Anos Noventa e o 'Novo Rural': Transformações Tecnológicas e Impactos sobre o Desenvolvimento Rural no Contexto da Agricultura Familiar no Rio Grande do Sul,* UFRGS/IFCH, Porto Alegre.
Graziano da Silva, J. (1991), 'Complexos Agroindustriais e Outros Complexos', *Reforma Agrária,* vol. 21(3), pp. 5-34.
IBGE, *Produção da Pecuária Municipal* – RS – 1990-1995 and 1997.
Magalhães, L. *et al.,* (1999), 'Competitividade e Políticas Públicas para o Agronegócio Brasileiro: Desafios e Perspectivas', *Indicadores Econômicos,* vol. 6(4), pp.196- 217.

Padula, C. *et al.*, (1999), 'A Cadeia de Suprimento no Setor Agroindustrial Leiteiro no Rio Grande do Sul: Uma Análise das Estratégias Empresariais Frente ao Mercosul', *Indicadores Econômicos,* vol. 6(4), pp. 218- 231.

Porto, M. (1997*)*, *Politizando a Tecnologia no Campo Brasileiro: Dimensões e Olhares*, Relume – Dumará, Rio de Janeiro.

Révillion, J. (2000), 'Estudo das Variáveis Relevantes na Adoção do Processamento UHT nas Agroindústrias de Laticínios do Estado do Rio Grande do Sul', UFRGS, Porto Alegre.

Tironi, E. and Lagos, R. (1991), 'Actores Sociales y Ajuste Estructural', *Revista de la CEPAL,* vol, 44, pp. 39-54.

Wilkinson, J. (1996), *O Estado, a Agroindústria e a Pequena Produção*, Hucitec, Bahia.

Chapter 10

Co-operatives to Companies:
The South African Wine Industry in the
Face of Globalization

Joachim Ewert

Introduction

Competition in international markets, in close conjunction with domestic
deregulation, has deepened the divide between 'winners' and 'losers' in the
South African wine industry – that is, between those growers and cellars
that are able to compete and sell their wine in international markets, and
those who cannot.

The potential losers are some co-operative wine cellars, their growers
and the workers employed on these farms. Shielded from competitive
pressures by a regulation regime for most of the Twentieth Century, cellars
and growers alike suddenly found themselves face-to-face with the harsh
laws of the market when the system was dismantled almost overnight
(between the late 1980s and the early-1990s). No longer able to rely on a
guaranteed minimum price for grapes and faced by cheaper imports, most
cellars responded to the new environment as best they could. At the
beginning of the new millennium it is clear that some of these responses
will not be enough to save them.

Following closely on the heels of deregulation, trade sanctions against
South Africa were lifted in 1993. At least theoretically, the reinsertion into
the global economy presented South African growers and cellars with
unprecedented opportunities. However, for many co-operative cellars,
exports have remained a mere hypothetical option. Not geared to the new
world of wine, many cellars were simply not able to take advantage of the
new opportunities. Still others, misled by reasonable, albeit short-term,
prices for their traditional output, decided to remain on the mass production
path.

Nevertheless, some actors in the South African wine industry are benefiting from entry into the new markets. They are the private cellars and co-operative wineries no longer wedded to the mass production paradigm. In any event, most private cellars were ready for the change when it came. Born from entrepreneurial zeal, and imbued with a business spirit, they have always spearheaded whatever innovation was introduced into the South African wine industry.

Faced with an enormous challenge, co-operative cellars have adopted a variety of responses in order to survive. Yet, despite these efforts, a large part of the cooperative sector faces the real threat of exit from the industry. Fifty per cent may no longer exist in 10 year's time. It is this fear of a massive shakeout – and its social consequences – that has prompted the established 'players' in the industry (excluding the private estate sector) to forge a strategy for a guided transformation, and not simply leave it to the market.

With the strategy having been formulated, it now awaits implementation. That, however, is not going to be easy. Massive capital requirements needed to finance new plantings and equipment is one obstacle standing in the way of a successful transformation. Changing the organizational culture of co-operatives and developing marketing expertise may even be more difficult to overcome.

If the co-operatives don't succeed, a shakeout is inevitable, and with it the loss of livelihoods. In the wake of increasing concentration and rationalization, the growth in employment over the last few years may yet be reversed, adding more people to a landless and jobless rural proletariat eking out a living on the margins of the Cape winelands.

The World Wine Market: No Call for 'Ordinaire'

Given the small and rather stagnant domestic market, South Africa's growers and cellars have little choice other than to break into international markets if they want to survive, or indeed prosper. What the global market wants is not cheap, low quality, wines but medium-priced wines, preferably distinguished from the rest by region and brand. As *The Economist* (Rachman, 1999, p.111) put it in a recent survey of the international wine industry: 'the message is clear: the best place is at the upper end of the market, where growth is fastest and margins are fattest'.

Standard price wines have poor prospects in a world where consumption has fallen by a quarter since 1982. In France and Italy alone consumption has halved in the past 30 years. All over the world, amongst those who do consume wine, there is a trend towards buying better-quality wines (Rachman, 1999, p.111). The most lucrative price bracket is the range costing over £4 in Britain and seven dollars in the US. In the latter market, wines selling at more than seven dollars account for only 22 per cent of sales by volume, but almost 50 per cent of revenues (Rachman, 1999, p.111).

The most important international market is Western Europe where 70 per cent of the world's wines are produced and consumed. Within this region, Britain is the all-important market. Unlike the parochial market of France, where imports command less than 5 per cent of consumption, Britain is the world's biggest importer of wine. France still has the biggest share of the British market, although this has fallen from 30 per cent in 1993 to 27 per cent in 1999. New world wine countries like Australia, on the other hand, have made significant inroads into the UK market, pushing up their share from 8 per cent to 15 per cent over the same period.

South African wines commanded a share of 6 per cent in 1999 (*The Economist*, 1999, p.98), reflecting a 1,000 per cent increase in exports between 1993 and 1997, although this was from a very low base (Ewert *et al.*, 1998, p.16). The share in the British market represents approximately 40 per cent of value and 60 per cent of volume of total South African exports. In 1997, South African exporters of wine shared a total revenue of almost US$200 million (Rachman, 1999, p.100). At the moment, 36 co-operative and 155 private cellars export wine from South Africa.

After spectacular growth between 1994 and 1998, exports grew by only 5 per cent between 1998 and 1999. No longer aided by novelty value and other short-term advantages, further growth of market share now depends on quality – as reflected in 'value for money'. To rely on cheap labour and a weak currency for competitive advantage would be clearly undesirable.

Within the British market, the supermarkets are the most important outlets. They now have over 60 per cent of the wine market (the situation is similar in the Netherlands; in the US 45 per cent of wine is sold in supermarkets) (Rachman, 1999, p.107). Most of Britain's eight big supermarket chains source wine from South Africa. To obtain a sales contract with one of these is a much sought after prize in the Western Cape winelands, especially for a co-operative cellar. In order to please the supermarkets, some cellars have even agreed to subscribe to ethical labour codes, albeit grudgingly.

Although South African growers and cellars are starting to learn the rules of the new competition, life is not necessarily going to become any easier. In certain quarters of the industry, a world wine glut is being predicted for the near future. According to *The Economist* 'a huge quantity of...high-quality grapes is about to come to market' (Rachman, 1999, p.113). In this scenario, competition between wine-producing countries and individual producers is bound to intensify and it is likely to become even more imperative to establish 'branded' wines that stand out from the rest. Traditionalist farming styles and an entrenched co-operative culture, characteristic of large sections of the South African wine industry, will be a hindrance rather than an asset, in their struggle for survival.

Survival is not made any easier by the fact that globalization is a two-way street. In 1998 South Africa imported approximately 16 million litres of wine products (as compared to approximately 118 million litres of exports). Of these, natural or good wines were the biggest item, accounting for more than 8.25 million litres (SAWIS, 1999). While imported quality wine remains unaffordable for all but the wealthiest section of South African society, it is the import of distilling wine that is impacting the most on South African co-operative wineries, many of which have come to rely on this 'traditional' product over decades.

Cut Adrift: South African Growers and Co-operatives in a World of Deregulation

'Planting after the Price': The Response of Growers

The lifting of trade sanctions in 1993, following shortly upon the heels of the abolition of planting quotas two years earlier, presented South African growers and cellars with a host of new opportunities. At the beginning of the 1990s, the majority of vineyards were based upon high yielding, nondescript, cultivars utilized mainly for brandy, fortified and distilled wine production. Wine struggled to increase its share of the domestic market for alcoholic products, encountering stiff competition from the beer producers. Most of the wine produced by the co-operative cellars was still sold in bulk to the Dutch-founded Ko-operatiewe Wijnbouwers Vereniging van Zuid-Afrika (KWV) and a handful of wholesalers, with the latter having a quasi-monopoly in the liquor retail sector. To make matters worse, international sanctions on agricultural products were imposed on South Africa in 1986. Only the grape juice sector managed to survive, with (clandestine) exports reaching new heights at the end of the 1980s (Visser, 2000, pp.27-29).

With political transition in the air, deregulation gaining momentum and the lifting of sanctions immanent, the position of the wine industry started to improve markedly at the beginning of the 1990s. For instance, grape juice exports increased by a further 140 per cent during the period 1990-91. Exports of good wine increased by 46.2 per cent during 1991, although this was from a very low base and insignificant if measured in volume (Visser, 2000, p.29).

Even before the first democratic election in 1994, South African growers started to replace existing cultivars and to expand vineyards. Between 1990 and 1997, the absolute area under wine grapes expanded at an annual rate of 1 per cent, while the gross tons produced increased by 2 per cent per annum over the same period (Ewert *et al.*, 1998, p.25). In 1998 the total area under vines amounted to 111,272 hectares and the total tonnage pressed to 1,041,004 (SAWIS, 1999). The expansion was mainly a response to the real (and expected) increase in wine producer prices in the context of growing domestic and international demand. Between 1990 and 1997, the minimum good wine price increased by 208 per cent, the rebate brandy price by 287 per cent and the distilling wine price by 311 per cent (Ewert *et al.*, 1998, p.24).

Most of the new plantings involved the so-called 'big six' (also called 'noble' cultivars) – Cabernet Sauvignon, Merlot, Pinotage, Shiraz, Sauvignon Blanc and Chardonnay. During the period 1996-98 alone, a total of 5,184 hectares of noble cultivars were planted, with Pinotage and Chardonnay showing the biggest increases. By 1998, noble cultivars represented 25.7 per cent of all South African vineyards, compared to 13.8 per cent in 1991 (Visser, 2000, p.32). As a consequence, 'quality' wine production (as measured by the 'wine of origin' label) increased from 17 per cent in 1992 to approximately 30 per cent in 1998 (SAWIS, 1999). Red wine production increased from 10.9 per cent in 1992 to 16.3 per cent in 1999. In 1998 natural wine pressed from the six noble varieties represented 27 per cent of all natural wine exports (SAWIS, 1999). Of all natural wine produced in that year, 21.8 per cent was exported (SAWIS, 1999).

Although the 'staple', nondescript, cultivars continue to dominate South African vineyards, the shift towards noble grape production represents a significant investment on the part of growers, with replanting costs of existing vineyards amounting to approximately R40,000 per hectare including soil preparation, trellising and automatic irrigation (Ewert *et al.*, 1998, p.29). This does not include the cost of land. On average, acquiring one hectare of new land would add another R100,000 to establishment costs.

However, the comparatively rapid expansion of vineyards and the concomitant shift to noble cultivars, has not been matched by an equally effective transfer of new technology and techniques to the farm. In a survey of 104 farmers in 1997, Ewert *et al.*, (1998, p.35) found that a minority of growers had changed or adjusted their viticultural practices over the previous ten years. For instance, only 47 per cent and 18 per cent practiced trellising and 'canopy control', respectively. This suggests that farmers are either not open to innovation, or that innovations are not being filtered through to farmers. In the view of some local researchers, the ineffective transfer of innovations from the experimental situation to the farm constitutes a major obstacle in the transformation of South African viticulture.

High Volume and High Sugar Content: The Regulation Regime and its Legacy

Between 1917 and the mid-1990s, the South African wine industry was regulated by an elaborate system of statutory rules, of which planting quotas, a minimum price for grapes delivered, and mechanisms of surplus removal, were the key elements. Carefully constructed, the system decisively shaped the dominant features of the South African wine industry.

As a result, the industry became almost exclusively oriented towards the domestic market. Having neither the know-how, nor the confidence, to compete in Europe on a large scale, the industry decided to restrict itself to the domestic market. Even today, after a decade of change, 90 per cent of production is sold locally (SAWIS, 1999).

Secondly, farmers and their co-operatives sold most of their grapes and/or wine in bulk to the KWV and a handful of big producing wholesalers (such as Stellenbosch Farmers' Winery). Although producers were guaranteed a minimum price, periodic surplus production and a limited domestic market continued to favour the buyer well into the 1990s. Most importantly, it made the co-operative cellars increasingly dependent on the wholesalers and their retail chains.

Thirdly, in a situation where a highly predictable system shielded producers from market pressures, neither the wine makers nor their staff were required to develop special skills regarding sales, price negotiation or marketing. With the bulk of the grapes destined for distilling and the sale of wine taking care of itself, there was little need for the development of a business outlook.

Fourthly, the emphasis was on the production of high volumes of low quality grapes. The signal farmers received from co-operative managers was to push up quantity and sugar levels as far as possible. Concepts like 'terroir', 'super blocks' (of vineyards), 'canopy control'and so forth, were not part of grower discourse before the 1990s. As a result, less than 20 per cent of South Africa's wine production fell into the premium price category by the mid-1990s. The 'pool' system became the lynchpin of co-op organization: all grapes of a particular cultivar (say, Chenin Blanc) delivered to the co-operative would be pooled in the same tank and sold mostly in bulk. Members would be paid proportionally according to the number of tons delivered and the selling price realized for the pool as a whole. The furthest co-operative management would go was to distinguish between one or more classes of the same cultivar, usually determined at the stage of delivery according to the sugar:acid ratio and the overall condition of the grapes.

Lastly, control of the co-operative cellar was exercised by a Board of Directors, mostly composed of the older, 'experienced' farmers. Typically, the wine maker also acted as the manager, being in charge of the day-to-day running of the cellar. The main objective of the co-operative was to realize the highest possible return (so-called 'payouts') for its members. Given a system geared to low-quality, high-volume production, the emphasis on maximum payouts was perfectly rational. However, it had the result that investment in new equipment was sorely neglected. Replacing crushers and filters and adding tanks for the purpose of quality production required capital. As legislation did not permit co-operatives to source equity capital on the open market, increased investment would have meant lower payouts for the members. In any event, with a low domestic demand for quality wines and international markets remaining closed to local producers, few saw the need for innovation.

In the end, comparatively little innovation was introduced into the cellars, either technically or organizationally. A system biased towards mass production had little use for the kind of know-how generated in the laboratories of the region's research institutes. As a result, the co-operative wineries were poorly prepared for the new market-driven environment, which they were about to enter at the beginning of the 1990s.

From Co-operative to Business: Forging Strategies of Survival and Growth

Deregulation and access to international markets opened up a whole range of new options to cellars, creating the possibility of bypassing the KWV and local wholesalers for the first time. The question of importance is: how have the cellars responded to these opportunities?

In a case study of 9 co-operatives undertaken in 1997 (Ewert *et al.*, 1998), all the managers interviewed professed a commitment towards more 'quality', using an absolute increase in good wine production as a yardstick (see Table 10.1).

Table 10.1   **Good wine production (million litres) by selected co-operatives, 1990–1996**

	1990	1991	1992	1993	1994	1995	1996
Winery 1	56.7	48.4	47.5	36.0	65.3	73.1	79.9
Winery 2	32.0	29.0	18.0	22.0	25.0	30.6	36.3
Winery 3	21.0	27.0	31.6	29.5	*	39.0	50.3
Winery 4	23.2	25.9	26.0	35.6	63.1	69.8	66.7
Winery 5	*	*	87.8	93.0	*	92.2	96.8
Winery 6	47.8	71.7	81.2	73.0	81.7	72.6	85.0
Winery 7	*	*	*	*	*	*	*
Winery 8	55.4	55.4	61.1	62.0	70.4	68.9	80.0
Winery 9	67.5	75.0	72.0	82.5	79.3	80.1	87.2

*data either not available or not released for publication

However, good wine production, as a share of total output, varied significantly between the nine cellars (see Figure 10.1). For instance, winery 1 focused on good wine production (bottled), winery 2 more or less split its production between grape juice concentrate and good wine (bulk), whilst the output of winery 3 was even more diversified than the latter.

Prices in the domestic market, plus operating costs, were the key variables in shaping cellar production strategy. At the time of the study, the domestic market had registered an increase in the consumption of medium priced wines and a decline in the demand for standard price wines. At the same time, the demand for brandy had increased, further strengthening the rebate brandy price. In addition, the price for grape juice concentrate had doubled over the same period.

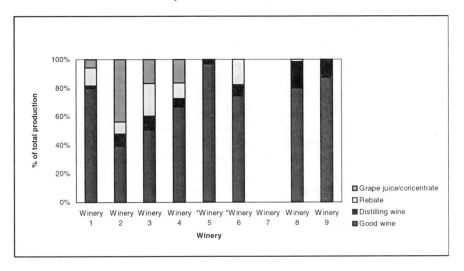

Figure 10.1 The utilization of the wine crop by selected co-operative cellars, 1996

As a result of these price movements, the co-operatives that had decided to forego the 'bottling route', were rewarded with relatively high returns from the market. In fact, winery 1, despite having added additional value, realized the lowest earnings of all nine cellars. In that case, higher prices realized in the market were offset by the high price of grapes sourced from outside the co-operative, plus the high costs of bottling (see Table 10.2).

Table 10.2 Relative profitability of wineries as measured by earnings before interest and tax (per cent), 1996

	Winery 1	Winery 2	Winery 3
Gross sales	100.0	100.0	100.0
Less cost of product sold	45.9	15.8	10.8
Gross profit	54.1	84.2	89.2
Less operating expenses	11.1	1.9	1.5
Earnings before interest and tax	**43.0**	**82.3**	**87.7**

Source: Ewert *et al.,* (1998, p.19)

Notwithstanding the short-term performance of Winery 1 in the example above, there is considerable agreement in industry circles that the 'added value' strategy is the appropriate course for the future, given the small domestic market and the trends in the global wine industry.

Moving from bulk production destined for the wholesaler, to bottled production earmarked for independent sales is one way in which co-operatives cellars in the South African wine industry are responding to a new environment. Changing their organizational structure and legal identity from cooperative to company is another dimension of the current transformation that is becoming increasingly popular. Providing management with more options, sourcing equity capital, changing the organizational culture and guarding against government interference are the main reasons behind this radical step (see Case Study A).

The company discussed in Case Study A represents one pole on a continuum of transformation in the South African wine industry today. The traditional co-operative represents the other. They could also be seen as two opposite categories of a typology (see Appendix 1). In reality though, most co-operatives are hybrids of the old and the new. Born out of necessity, they have devised arrangements that give them more flexibility, but stop short of far-reaching structural transformation. For instance, some cooperatives allow growers to operate outside the grape delivery agreement, but keep them on board as members. Others have added a marketing division to the existing co-op structure. A handful have gone a step further and established a marketing company with an overseas partner. The case study below illustrates some of the careful steps cooperatives have taken in their attempt to stay afloat and face a new future.

The co-operative in Case Study B, like the company in Case Study A, is almost certain to survive. The same cannot be said of some of the others. After an increase of 20 per cent in the mid-nineties, domestic sales of wine products declined again towards the end of the decade. Following upon the spectacular 'take-off' between 1994 and 1998, exports grew by only 5 per cent between 1998 and 1999. By the end of 1998 the novelty of South African wines had worn off. Competition became stiffer as overseas buyers increasingly demanded value for money. The dubious quality of some of South Africa's initial exports, combined with the ineptitude of local wineries and an inadequate logistical system, once again started to compromize South Africa's position in international markets. As international success started to wane and stocks started to accumulate, producers turned to the local market to rid themselves of their produce, including premium quality wine. The result was the biggest surplus in the history of the South African wine industry.

Case Study A: Former Co-operative, Now Company

Exemplifying the 'company type' is a new company which resulted from a merger between four former co-operatives. In this company, shares can be traded freely. In the course of the transformation, former co-op members were given the option to stay on as shareholders or to sell off their equity. At the same time, non-farmers can acquire shares in the company. The cellar buys grapes from erstwhile members/shareholders, but also from other suppliers. Producing shareholders, in turn, are not obliged to sell to the cellar, but are free to sell to the highest bidder.

As a matter of policy, this cellar negotiates the price for grapes anew every year. In a market-driven environment, there is no guarantee of a linear increase in producer revenue, since the latter depends mainly on the price the grapes fetch under current market conditions. However, growers get paid for their grapes almost immediately after delivery, and not in two or three instalments, sometimes stretching over 18 months, as is the case for the classical co-operative.

At the operational level, this newly formed company cellar has added extra tanks and other equipment to its existing production technology, enabling them to produce 'batches' of different quality wines. In the process of re-tooling, special attention was paid to the development of a red wine production 'line'. In a joint venture with other cellars (including at least one other co-operative), the company has installed a bottling line on its own premises, making it independent from specialized bottling companies.

At this cellar, technological change has gone hand in hand with the professionalization of cellar management. This includes the recruitment of 'flying wine makers', as some traditional South African wine makers have found it difficult to make the switch to quality production. It also involves the appointment of cellar managers and directors with a commercial/corporate background, including some with no previous experience of the wine industry. The company has also appointed a vineyard expert, who visits producers on a regular basis, giving advice and identifying the best vineyards (prime or 'super blocks'). Making the transformation complete, the management board has appointed professionals to take control of the company's marketing. At a psychological level, the management team has attempted to banish the word 'co-operative' from the vocabulary of the staff and are unflagging in their efforts to propagate the benefits of being a 'company'.

Case Study B: Co-operative Winery in Transition

One co-operative in the Stellenbosch region illustrates the process of adaptation. In the mid-1990s, the cellar and its 21 members invested in new equipment, not procuring state of the art technology, but 'only what was necessary' to produce better quality wines. One reason for keeping the capital outlay within limits, was not to reduce payouts to the point where they had no money left for replantings. Replantings and investment paid off in the form of top awards for certain red wines at South Africa's premium wine show (the Veritas Awards) towards the end of the 1990s. As part of the quest for quality, the cellar requires its members to plant only noble cultivars and to reduce the yield per hectare. It has also introduced 'block grading' in order to identify superblocks for the making of superior wines. At the same time, the cellar allows its members to deliver grapes directly to a producing wholesaler, as long as this is regulated contractually between co-op and wholesaler. Also, the co-op has a special relationship with one non-member, for whom it produces a private bin, thereby utilizing capacity to the full.

However, the co-operative continues to sell most of its wine in bulk to one of the biggest wholesalers in the industry, in accordance with an 'evergreen' (long-term) contract. This is despite the fact that the co-operative is part-owner of a bottling plant. At this stage, the co-operative bottles relatively little of its wine, most of which is exported to the Netherlands. The co-operative is on the verge of a further stage of technological renewal. It hopes to forge a partnership with a British supermarket, thereby acquiring a much needed capital infusion, using existing marketing expertise and increasing its export share, all at the one time. At this early stage, taking care of its own marketing and sales – both locally and abroad – is being regarded as too risky.

Wrong investment decisions, path dependency, high interest rates and a co-operative culture that was difficult to extinguish, have brought a number of co-operative cellars to the brink of a merger or even exit from the industry. Left to their own devices, they simply will not survive. It is for this reason, that the traditional 'establishment' of the industry, albeit under a new name, commissioned a strategic study into the future of the wine sector. Operating under the title 'Vision2020 Creating a Future for the South African Wine Industry', the strategy team was given a threefold brief: to undertake a so-called 'mess analysis', spell out different future scenarios, and facilitate a transformation.

After a series of workshops, the team produced a 'strategic agenda', highlighting the vital steps the industry needs to take if it wants to survive. Topping the list is the imperative to 'achieve international competitiveness' through 'entrepreneurial [and] highly innovative' action. Strong emphasis is also placed on the required shift from a 'production-driven to a market-driven wine industry', including the development of a 'Brand South Africa'. In this regard, the study team does not believe that it is necessary to 'improve' the varietal composition of the South African wine industry. Instead, it thinks that 'careful attention to the detail of producing quality wine grapes from a specific terroir and vineyard is the first building block...for brand development' (Winetech Vision2020, 2000, pp.1-3).

In the strategic paper there is not a single reference to any sort of re-regulation. At the same time the study emphasizes the need for a 'governance system' in order to prevent it 'muddling along'. The effective long-term development of the industry, the study argues, '... will in future depend on the quality of market- and industry-information that is available, (from centralist statutory controls to decentralized information-bonded development)'. Whether any institution is going to act as a clearing house or disseminator of information, is not clear at this stage.

In the meantime, economists in the employ of 'VinPro' (formerly KWV Consulting Services, now privatized) are taking the Vision2020 'gospel' to the road. Working from the assumption that the existing surpluses are 'not a production problem but a management problem', they are trying to transmit management skills to cellar staff, many of whom were originally trained as wine makers. However, for some cooperatives this new knowledge appears to have come too late. At the time of writing, at least two co-operative cellars were up for sale. Almost certainly, more will follow.

What About the Workers? The Impact on Employment and Livelihoods

Deregulation and the exposure of the South African wine industry to international markets are not threatening everybody's livelihood. For instance, a study published in 1998 (Ewert and Hamman, 1999, p.210) found that regular employment in the industry had increased by 7 per cent over the period 1994-97 (the period immediately following upon the lifting of sanctions and the 'take-off' of exports). At the time, the industry provided jobs to approximately 50,000 people. The same number of jobs is created during the harvest season in February and March each year. If the dependants of regular workers are counted in, the industry provides livelihoods for at least 200,000 people. These are important figures in a country where jobless growth is the norm and where unemployment is estimated to be standing between 23-34 per cent of the economically active population (SA Institute of Race Relations, 1999, p.247).

Whether regular employment has decreased in the wake of lower domestic sales and exports since 1998, is difficult to say. No recent statistics are available. Also, if aggregate employment has declined, this is not necessarily the result of lower sales, whether in the domestic or the international market. A smaller labour force, plus the observed casualization of labour, may rather be attributed to a combination of rationalization, mechanization and new labour laws (Ewert and Hamman, 1999).

Whatever the case may be, the wages of those fortunate enough to retain their regular employment have not increased radically in the new, post-sanction, era of the South African wine industry. Currently it is estimated that men earn R165 cash (or approximately £15) for a 45 hour working week. This reflects an annual increase of 12 per cent since the mid 1990s. However, if inflation (especially for foodstuffs) is taken into account, real incomes have hardly improved over the last six years. Although no recent survey data are available, there is no reason to believe that housing conditions and services on farms have improved radically in recent times. Thus, regarding income and living conditions, workers in the South African wine industry have benefited little from the minor export boom. In the face of weak state capacity to enforce labour legislation and the extremely low level of unionization amongst farm workers, it is unlikely that workers will be able to gain a bigger share of the industry's revenue in the near future.

Conclusion: Deregulation of the South African Wine Industry – Who Benefits?

Deregulation of the industry has put in jeopardy the continued survival of a significant percentage of the co-operative sector and all the people who depend on it for their livelihood. No longer able to rely on state protection and facing a saturated domestic market, exports appear to represent the only way out of the crisis. However, this is only in theory. In fact, it is most of the marginal cellars that find themselves in a classical double squeeze: falling revenue on the one hand, and huge capital requirements for the purpose of restructuring on the other. Given that government assistance, in the form of subsidies or otherwise, is out of the question, and banks that are not willing to extend any further loans, bankruptcy may only be a matter of time for a number of co-operatives. No amount of management advice, coming from 'Vision2020' or elsewhere, will avert closure. Old established cooperatives are certain to disappear in the wake of mergers or acquisitions by private investors.

Interestingly enough, the immanent demise of these co-operatives is not particularly mourned – except by those directly affected by it. Certain sections of the industry, especially the private cellar constituency, as well as members of the public, have felt for some time that excessive regulation and government protection has unduly sheltered thousands of mediocre farmers and a horde of bureaucratic officials, charged with the detailed policing of the regulation regime. It is argued that innovation in the industry has been held back on all fronts. Those opposed to regulation received confirmation for their views when South Africa re-entered the global wine market.

Less than a decade after the industry entered the international market, the majority of South African cellars – especially the private ones – have a stake in the exporting and are benefiting from it, either because they were well positioned when sanctions fell away, or because of successful adaptation to international price regimes. They are not only benefiting from revenues in hard currencies, but also from the exposure to new ideas, innovative practices and marketing networks. Naturally, they are unequivocally in favour of deregulation. At the same time, they have little sympathy for the co-operatives which find themselves unable to adjust.

Access to international markets has been the driving force behind the extensive new plantings in the industry since the early 1990s. This, in turn, has been responsible for significant growth of jobs in the industry. Whether this trend will continue remains to be seen. For the moment, new plantings have been put on hold. Also, rationalization of labour arrangements, and mechanization, are expected to gather further momentum. Together with the predicted closure of some cooperatives, these tendencies will act as brake on further employment. The creation of more jobs will largely depend on the ability to capture more ground in the export market.

History has taught us that workers do not automatically share in the gains accruing to a growing sector. The South African wine industry is no exception. So far, globalization has brought them little improvement, either in the form of wages or living conditions. Nor has the new, democratic state been of much assistance. Progress for the workers will, it seems, once more depend on their own struggles.

References

Dreyfus, F., Martin, C. and Ewert, J. (2000), *The Socio-Economic Transformation of Viticulture – A Comparison Between Languedoc-Rossillion, France and the Western Cape, South Africa*, Proposal for a Research Programme, unpublished.

Ewert, J., Hamman, J., Tregurtha, N., Vink, N., Visser, C. and Williams, G. (1998), *State and Market, Labour and Land – the South African Wine Industry in Transition*, Research Report, Department of Sociology and Department of Agricultural Economics, University of Stellenbosch, South Africa.

Ewert, J. and Hamman, J. (1999), 'Why Paternalism Survives: Globalization, Democratization and Labour on South African Wine Farms', *Sociologia Ruralis*, vol. 39(2), April.

Rachman, G. (1999), 'The Globe in a Glass: A Survey of Wine', *The Economist*, 8 December, pp. 97-115.

South African Institute of Race Relations (1999), *South Africa Survey 1999/2000* (nd).

South African Wine Industry Statistics (SAWIS)(1999), *South African Wine Industry Statistics 1999*, see website: http://landbou.com/sawis/

Visser, C. (2000), *South African Co-Operative Wineries – History and Current Status*, unpublished manuscript.

Winetech Vision2020 (2000), *Strategic Agenda for Discussion by Study Groups*, mimeo.

Appendix 1   Typology of wine cellars

Type		Traditional Regime	Quality Regime
Structure	Status	Co-operative	Company (merger or not)
	Skills	Wine maker + manager = same person	Recruitment: 'flying' wine maker, professionals for marketing/sale, vineyard expert Appointment of cellar manager + director
Wine making	Type of wine	Distilling and rebate wine > 50 per cent mostly white	Good wine > 50 per cent; increasingly red
	Quantity production	Higher	Lower
	Processing mode	'Pool system'	Separate tanks
	Recent investment	No	Yes: tank, crusher, filter...
Economy	Production costs	Lower	Higher
	Earnings	Higher	Lower
	Sale price	Lower	Higher
	Selling rate	Lower	Higher

Relations with members	Type of members/ evolution	Growers, departure of 'quality' members	New shareholders (non-members before)
	Accounting system	2-3 instalments over 1 year or more	1 instalment over less than 1 year
	Accounting price	Lower	Higher
	Grape differentiation	None	Several classes/quality => differential accounting
	Advice/ evaluation	None	Type of cultivars, quantity required, canopy control, harvest date, work conditions, block grading,
Relations with buyers	Market	Mainly domestic	Domestic + international
	Marketing	Bulk only	Bottled
	Type of buyers	Mainly wholesalers	Wholesalers, direct sale (retail), international buyers, supermarkets (national and international)
	Brand image	Standard label	Quality ('terroir' and wine making conditions) and ideological images (social conditions of workers)
Relations with research		None	Information seeking re: innovations
Recent innovations		None	Technical, institutional and/or organizational

Source: Dreyfus *et al.*, (2000).

PART III
COMMUNITIES AND RESISTANCE
IN A GLOBALIZING WORLD

Chapter 11

Two Roads to the Global Village: A Comparison of How a Coastal and a Mountain Region of Norway Have Found Strategies to Cope with Globalization

Reidar Almås

Introduction

This paper will outline two different socio-economic regional strategies to meet with the challenges that have arisen from market and political forces over the last 20-30 years. The rise of globalization as a concept and process in development policies goes hand in hand with the economic liberalization and deregulation policies of the 1980s and 1990s. As stated by Philip McMichael (1996, p.9) the 'development debate cannot be adequately understood or resolved without a global perspective'.

'Regional strategies' may, however, give a false impression. The two 'strategies' were formed as a macro sum of all strategies followed by local actors living in these regions to cope with changes in markets and policies. The inland region of Røros in mid-Norway was formerly based on mining and mountain farming, while the resource base in the island region of Hitra-Frøya was fishing and coastal farming. In the 1970s both regions experienced a deep economic crisis because of a decline in their traditional way of making a living.[1]

In the mountain region, the Røros copper mines were closed down in 1977 after 333 years of active mining, being at one time the most important copper mine in Norway. Mountain farming with mostly small plots also underwent a radical structural rationalization in the 1960s and 1970s. During the 1960s and 1970s the primary industries in the coastal region also underwent major restructuring. The traditional coastal fishing system where one man used one boat to make a living for his household was not economically and socially attractive any longer for young workers. Coastal farming also underwent a large decline, partly because the combination of farming and fishing – which was necessary for those having small plots – almost disappeared.

In 20 years, from 1951 to 1971 the coastal region lost 1,900 of its 12,200 inhabitants. And the mountain region in the same period lost 1,200 of its 1951 population of 13,000. But while the population decline in the coastal region went on until recently, the mountain region managed to stabilize its population in the mid 1970s.[2] Today, the mountain region still has almost 12,000 inhabitants, while the coastal region has just above 8,000. Why this difference? Do we find the answer in different mentalities, in national policies, or in differences in the economy and natural resource base? Or do we find the explanation in the way the two regions are influenced by the process of market globalization?

Today, the coastal region is being integrated into the international fish markets through a highly competitive fishing and fish farming industry. Big companies are integrating forward and backward along the food chain, and politicians are saying: look to Hitra and Frøya for the future. The North Sea based oil industry is also making its impact. Near Hitra, at Tjeldbergodden, North Sea oil is taken ashore and processed in high-tech petrochemical industries. A technology research company (SINTEF, 1999) has pointed to mariculture and fish-based coastal industries as the booming industry of this century, which will become the most important Norwegian export industry when the oil is gone. Tourists are flocking to the coastal region. Will this promising future materialize, or will population decline continue?

In the mountain region, mining has been abandoned completely and the mining town of Røros has been converted into a regional tourist centre based on artefacts from mining culture, mountain farming and handicraft. Green tourism and eco-tourism are thriving on the close links to small-scale mountain agriculture. However, this small farming culture is struggling hard to survive under the pressure from deregulation and WTO demands. In order to do that, it is building coalitions with consumer and environmental groups in Norway. A new generic label concept has been developed to market food from the region: 'Food from the mountain region' *(Mat fra fjellregionen)*. At the same time, one of the most successful and high-tech industrial producers of office chairs in Europe is located at Røros, exporting ergonomic chairs all over the world.

The question I ask is why was the mountain region able to halt the economic decline in the 1970s, and adapt to globalization so successfully, while the coastal development was so different? Their reactions and strategies may be explained in different ways, as mainly connected to their resource base, their former links to the world market or by different cultural and political orientations. Is it at all possible to orchestrate regional political strategies to cope with globalization, or do we have to wait for the invisible hand, making some regions socially and economically viable, while waiting for others to die? I argue that there is more than one road to globalization, and that the end of the story is not necessarily a disaster for all rural communities.

The Mountain Region of Røros

The Røros region is situated in the eastern part of middle of Norway, close to the Swedish border. When you drive up the valleys from the north or the south, you get on to a rather flat, treeless mountain plain 600 meters above sea level. The pine trees were cut down and used in the mining industries, and never regenerated. As I define it, the mountain region comprises the municipalities of Røros (pop. 5,565) and Holtålen (2,229) in South Trøndelag county and Tolga (1,812) and Os (2,168) in Hedmark county. All these municipalities were within the circumference of the Røros Copper Mines, which from 1644 to 1977 defined this mountain region as a centre of mining activity, both culturally and economically. When the last mines were closed in 1977 (because of depleted copper resources and falling profits) the region was forced to redefine itself.

Today we know that this redefinition was a success story, but in 1977 most people in the town of Røros thought they could only look forward to living in a ghost town, which so often is the case in the closure of mining towns all over the world. How was it possible, during the age of globalization in the 1980s and 1990s, to redefine the Røros region's role, economically and culturally, in a way that may be called a successful symbiosis of old and new traits? How was it possible to avoid big internal political and social fights? And what can we learn from that, as far as rural planning and policy is concerned?

When mining ceased in the Røros region, the main socio-political actors were looking for alternatives and new beginnings. Politically, this was and still is a 'red' area with a collectivist mentality, where the Norwegian Labor parties have had a stronghold since the early days of industrialization. The region has always sent radical and socialist politicians to Parliament, among them the famous author Johan Falkberget (1879-1967), who was a miner himself up to age 27. In his writings Falkberget in many ways expressed and defined the ethos of the Røros region in the early 1900s:

collectivist-oriented social radicalism mixed with strong hardworking Protestant ethic and cultural conservatism.

This rather curious blend of mentalities, which to my knowledge is not to be found in other mining towns of Norway, appears to have contributed to the successful strategies chosen out of the ghost town possibilities. Ideologically, the labour movement in the Røros region could have chosen a militant struggle against the mining company during the 1970s. Instead, they chose to work for a new beginning together with other social partners. Mayor Ole J. Kverneng (1952-63, 1964-67) from the Labor Party made an important contribution here, forging the foundation of a social contract between employers and employees, which is still valid at Røros. In doing so, he followed the ideology of reformist social democracy, which is also to be found in the writings of Falkberget. In these writings there is both a class struggle and a mental struggle between good and evil in people's minds. But there is always a possibility that the good forces in the minds of the upper class people may join hands with the good forces in working class minds. His heroes are women and 'intellectual' outsiders who come into the mining community with new ideas, often against the ruling ideas.

In addition to the labour class working in the mines, the mountain region has for a long time had a large group of farming smallholders earning most of their incomes outside agriculture, as part-time wage labourers. Actually, most of the mining households also had tiny farms, even those living in town. This system of subsistence mining economy kept down the costs of living and consequently the wages were kept lower than elsewhere in the mining industry. Hunting and fishing also contributed to household production. However, this natural resource-based, part-time adaptation was not for the mining workers only. Especially in the municipality of Holtålen, the largest groups of farms belonged to this subsistence system where the father worked far away from home in construction, on the railway, or in other industries. This system has prevailed up to this day, and explains why 72 per cent of the farms still have below 5 hectares of arable land in Holtålen. In Røros 62 per cent of farms are this size. Today, most smallholders with such micro-farms have their main employment in the modern construction industries or, more and more, in the growing service and public sectors. But their ethos and cultural mind set is still shaped by the small-holding proletarianism created before the Second World War.

The smallholders in the Røros regions were also more collectively oriented than most farmers in Norway. In Rausjødalen in Tolga the first large-scale co-operative cheese industry was established by 20 farmers in 1855. This co-operative mentality has lasted to the present, and in the

1960s and 1970s small farmers established several group farms in the mountain region communities (Almås 1980, p.274). The smallholders in the mountain region, by choosing to modernize their farms jointly, and by renting their neighbours' land rather purchasing it, did not offend the collectivistic tradition of the area. Through conversion from smallholder plot farming to group farming, the social fabric of the community was to a large extent preserved (Almås, 1980, p.205).

Figure 11.1   Hitra-Frøya Region and Røros Region, Norway

We should not forget, however, that agriculture in the Røros region has mostly been organized along family-farming lines – similar to farms in other mountain areas of Norway. The short summer season and the high altitude have always reminded the mountain farmers of how ecologically marginal their farming is. The mentality of the region has also always been known for a modest and unassuming attitude towards nature and life. This mentality may explain why organic farming today is very widespread in the mountain region. Not only among farmers who are classed as organic farms, but also among those who run their farms with conventional methods, there is a very strong ecological consciousness. The present efforts to build the generic label of 'food from the mountain region', partly stems from this eco-consciousness.

Eventually tourism became a very important industry in the Røros region. Based on the preservation of the 300 year old wooden buildings in the Røros town itself, tourism, as an industry, employs 12 per cent of the economically active population. Several unique museums in town as well as in the neighbouring communities – including a magnificent church from the 1780s – as well as production and sale of arts and crafts draws crowds of tourists from all over Europe (Daugstad *et al.*, 1999). A prominent place on UNESCO's World Heritage List of the world's most valuable attractions has been won as a result of co-operation between private and public interests in the town and in the region. Several of the mines in the Røros region have been converted into museums, and are also used for concerts and theatre performances. The special cultural atmosphere of the town has attracted several groups of artists to the area, and their art works, together with the special foods, add to the tourist experience of locality and regionality.

The Island Region of Hitra-Frøya

The coastal region, which consists of two municipalities, Hitra and Frøya, has a very different economic history, both being closely bound up with the fishing industry, but with Hitra also having an important agricultural sector. Agricultural land is scarce on both islands, but Frøya is even more rugged and broken up by rocks and cliffs. Traditionally, most people have been living from fishing and related industries, and a boat was necessary to reach the mainland. Fishing for subsistence was practised in almost every household until the 1970s. A very familiar household adaptation was fishing in combination with a small farm (less than 2 hectares). Hard weather and risky boats made life on the island harsh, but the protein resource from the sea was a guarantee for survival, even in times of hunger.

People in this region have had extended contacts up and down the Norwegian coast through fishing and sea transport, as well as to foreign countries through employment in the merchant fleet. This seafaring tradition has resulted in a rather cosmopolitan culture. The islanders have also had to rely very much on themselves without governmental help in many cases, which is exhibited in the culture as a strong streak of individualism. The owners of fish factories and big boats have had much power, which we find in the political culture as paternalism and a tendency to side and vote with one's patron in conflict situations.

Recently, the two islands have been linked to the mainland by an underwater tunnel. Hitra was linked to the mainland by the world's deepest road tunnel (264 metres under the sea surface) in 1994. The last tunnel connecting the outer island, Frøya, to Hitra was opened in June 2000. After a steady decline in population since the Second World War, it appears to have stabilized at around 8,000 in the costal region. Why did this population decline stop?

Hitra has a population of 4,060, and in addition to the main island it consists of 3,300 smaller islands. One-fifth of the economically active population work in farming and fishing, 15 per cent are employed in industrial production, while one third are employed in the public services. On the island are 700 freshwater fishing lakes. Northern Europe's largest population of stag roam freely on the island. A tourist business is developing quickly, particularly since the time the island was connected to the mainland. Hitra is a very popular place to have a cabin – a favourite recreational option in Norway. The waters off the coast of Hitra and Frøya are rich with fish, and many tourists come here to join the local captains on fishing trips.

Hitra's farmers were known nationwide early in the television age when they started a tax strike in 1975 (Almås, 1978). At the time, farmers' incomes were relatively low, the restructuring of farming was rapid, and rural areas like Hitra were quickly depopulated. The strike was widely covered by media and the small farmers won sympathy in wide circles. The Minister of Agriculture had to go to Hitra and talk with the farmers in a public debate. As a result, later the same year the Parliament made a historic decision that farmers' incomes within six years should reach the same level per man-year as an average yearly income of an industrial worker. The farmers were also 'lifted into' the welfare state, getting health benefits and a farm relief service financed by the State.

The primary industry on Hitra, however, is fishing and fish related industries. When fish farming started in the early 1970s, Hitra was the most important entrepreneurial location. Family based to begin with, fish farming grew out of the traditional fishing and fish-processing milieu, and became rather quickly a backbone of the local economy. Initially the industry grew very fast, with few problems. However, in the late 1980s and early 1990s, many fish farms suffered from disease problems and the industry also experienced problems of overproduction and dumping. Efforts to solve overproduction problems with co-operative and government intervention to buy and store frozen salmon failed, and bankruptcies and major restructuring were the result. Today fish farming is dominated by some major firms, which have bought up many smaller concessions. These firms integrate both backward into feed production and forward into processing, marketing and sales abroad. The first major transnational buy-up occurred recently, when the Holland based multinational feed company Nutreco purchased the fish farming company Hydro Seafood.

Frøya has a population of 4,108 and consists of one main island and several smaller populated islands with fishing villages like Mausundvær and Sula. Fifteen per cent of the economically active population are employed in fishing, 25 per cent are employed in industrial production, mainly in fish processing, and one third are in the public services. Frøya is even more dependent on fishing than is Hitra, and its fish industry has also to a large degree remained within deep-sea fishing with seagoing trawlers. Agriculture is basically part-time farming, mainly meat production. North in the municipality lies the landscape protection area, Froan, which is well known for several endangered species of plants and birds. Communication has always been a problem and more so than with Hitra, because Frøya is further out to sea and because more of the islands were populated. However, there has been an internal centralization over the last 50 years, leaving some islands and communities depopulated. Frøya had a competitive advantage earlier when closeness to the fish banks was important, but this has almost disappeared with the introduction of trawlers with processing capacity on board. However, there are good fjords and inlets for fish and shellfish farming, which adds a competitive advantage for these future ways of cultivating the sea.

Like Hitra, Frøya has been integrated into the global economy through the fish industry and fish farming. This integration has increased recently, and this process will be speeded up by the new mainland connection. The fish processing industry has experienced difficulty finding a stable labour force, and has been obliged to recruit workers from neighbouring countries and among political refugees to Norway. Working in the fish industry seems to be of low status for the population of the fishing villages, and the lack of local interest together with the inability to find competent recruits seem to be the Achilles' heel of those coastal communities. Manual labour generally and especially in the provincial based industries, seem to be of low-status among Norwegian youth, even though it is rather well paid.

Global Impacts

Where and when does globalization enter these regional scenes? It is obvious that there have been global impacts throughout the course of the histories of these regions. Just to take two examples: when Norway had to change its alcohol policy in 1926 after a short period of prohibition (1919-26), much of the pressure to do so came from fishing districts like Hitra and Frøya. Their economies at the time were completely dependent upon export of dried cod to Portugal and Spain. When these two countries started a boycott of Norwegian fish exports because they could not sell their sherry and port wines to Norway, workers in the fishing areas protested loudly and the referendum majority changed from yes, to no, to prohibition in a few years.

The other example may be taken from the mountain areas, where they relied heavily on copper export markets and copper prices. These markets have always oscillated with export trends, especially because of wars, as copper has been used in weapons and ammunition since powder warfare appeared. Standards of living in the mountain region therefore were going up and down with war and peace. Wars all over the world have provided labour and income to the Røros region, not just to the miners, but also to the crofters driving coal and copper, as well as to the merchants selling goods and to the farmers feeding people in the Røros area with food and fibre.

How, then, are international trends and forces influencing the study areas in the new era of globalization? Much of it has to do with the speed and areas of impact. Just to take one example: salmon is slaughtered in the morning at Hitra, packed, iced and sent to the Trondheim Airport by freezer lorries. There it is loaded onto a large Iljusjin airplane from the privatized Russian company, Aeroflot, which takes the salmon to Tokyo, Japan, where it is consumed in sushi dishes the next evening. This example illustrates both speed and scope of Norway's connection with the global marketplace. What happens at one location can be reached from – and have an impact upon – anywhere in the world, which brings both threats and possibilities. And the speed of impact is reduced from decades and years to hours and days. *Globalization means accelerating internationalization in all spheres of life to all corners of the world.*

To take another example from the outmost islands in Frøya. At Froan, there is a conflict between environmental authorities and the salmon production company Salmar – the company wanting to fish farm in the Froan landscape protection area. If Salmar proceeds and develops fish farming facilities, it will be very susceptible to international boycotts from environmental groups. Because the company relies on global markets where environmental groups have significant powers of persuasion, the company cannot afford to overlook environmental protests.

What can we learn from this? First of all: it is possible for communities and regions to simply adapt to outside forces, rather than to shape those forces. As Putnam (1994) has shown with case studies from Emilia-Romagna, Italy, the type of political culture and the level of participation strongly influence the regional adaptation and outcome of pressure from market forces and globalization. Duncan (1992) has shown that communities with a high level of public participation and a vigorous civic society fare better than those communities that are not as well endowed with social capital.

Actor Networks, Industrial Districts and Social Capital

What theoretical concepts do we need to grasp what happened in the mountain region and what did not happen in the coastal region? Concepts from actor-network theory, although developed to study technological development, may give us some clues. The actor-network theory originates in an attempt to unite a micro-sociological perspective with a structural perspective. The source of inspiration for this theory of technological development by following the actors, is the French writer, Bruno Latour (1987). Latour focuses on scientists and engineers in action and asks questions like: How do inventions and innovations come about? How are machines and production processes modified during construction and adaptation of prototypes? Who participates, who contacts whom? Who will be affected and how do personal actors and institutions react?

Latour's methodological scheme has inspired geographers and agricultural sociologists to use a network perspective in viewing the development of production chains in agriculture (Ward and Almås, 1995). By analysing the growth of the Canadian soya complex, Busch (1994) shows how various social groups have persistently participated in the necessary shaping of the technologies and the organization of production and sales. Busch claims that soya researchers do double work: they are responsible for developing soya cultivation both biologically and technically, but at the same time they also work to create social acceptance for the system of technology and production. This is true not only for researchers, but also for farmers, environmental activists, consumers, politicians and other relevant social groups.

This theory has been extended by English geographers and sociologists in studies of actors in networks (see Marsden *et al.*, 1993). By analysing formal and informal contacts and bonds between actors, it is possible to see how these are woven together in networks. These may vary in form, such as producer/customer, producer/wholesaler, technological relationships, advisory relationships, and political and organizational types of relationships. In order to accentuate the importance of the wider context, these researchers have suggested calling the theory '*actors in context*' (Marsden, 1992; Marsden *et al.*, 1993). The theory of actors in context is well suited for developing models of the way in which various types of networks are created in business development. When we look at the relationships between key politicians, business leaders, and a few other persons in the Røros region, we find that these were absolutely essential in facilitating take-off in the 1970s and 1980s.

From a structural viewpoint, the theory of industrial districts (Pyke *et al.*, 1990) can also be of some help. The theory of industrial districts has its origins in the literature about flexible specialization and institutional economy. Such industrial districts have been studied closely in Emila-Romagna, Italy (Putnam, 1994) and Småland, Sweden (Johannisson, 1994). There are several definitions of industrial district, but Amin (1994) has identified four key characteristics: (i) division of labour and cooperation between businesses, (ii) a structured form of socialization and training, (iii) a local industrial atmosphere or environment, and (iv) institutional 'density'. Institutional density means incidence of private or public service or advisory organs, commonly owned businesses, and various cooperating organizations. The density of this network increases with an increase in the number and strength of these institutions (Amin, 1994). It is a good reason to believe that the Røros region became an integrated industrial district in the late 1980s or early 1990s.

Studies of industrial districts, all with their unique characteristics, shows that a differentiation is taking place (McMichael, 1996). It is therefore necessary to view the idea of globalization in a new light. The explanation is not a single form of logic that integrates a large, global system, but many forms of logic that only can be explained in the context of local conditions.

Asheim (1991) has pointed out that it would be difficult to produce industrial districts through economic development or efforts to stimulate growth. The districts are often the result of a long industrial and cultural development where part of the key to success lies in a combination of functional and territorial integration.

In order to study the balance between the local and the global, theories of 'embeddedness', civil society, and social capital also have explanatory possibilities. The theory of 'embeddedness' has many roots (see, for example, Amin and Thrift, 1994), but the one having the most relevance for analysis of rural development is that of Granovetter (1985) which asserts that economic behaviour is strongly embedded in interpersonal networks of social relations. The confidence that is required for taking necessary risks in economic transactions, originates in those social relationships that surround an economic action. In local societies, such relationships may be friendships, family relations, or neighbourhoods. A common set of norms and rules reduces the risk in economic transactions. In an extended context, it is the organizations, networks and institutions – in other words the civil society – that 'embeds' the economy.

As mentioned above, Putnam (1994) shows how traditions in the civil society are a key to understanding economic growth and success. Through examples, especially from Italy, he shows that regions that have active choirs, soccer teams and people who participate in organizations are also well developed economically. The social and cultural capital (Bourdieu, 1979; 1993) that is created through these horizontal networks partially explains why some regions develop well, while others fall behind. According to Putnam, in addition to markets and money, the most important discovery in economic history is credit. 'Impersonal' credit assumes confidence in the debtor returning money. The word 'credit' comes from the Italian word 'to believe in', which Putnam (1994) regards as a linguistic illustration of how business, in order to function well, must be surrounded by a civil society of trust.

Why Was the Røros Readjustment a Success?

Politicians and analysts of the 1970s could not possibly foresee the success of the Røros region. Norwegian farm policies in the late 1970s were quite generous to areas like Røros. New start-up businesses also were supported, and local regional efforts were quite vigorous. There appear to have been six factors contributing to the successful and 'smooth' restructuring of the Røros region's economy and its cultural redefinition. These were:

- The strong idea of co-operation in the Røros labour movement especially and in the mountain region generally – a major social capital asset that was well used;
- The scaling up of Norwegian agriculture in the 1970s, which meant a stabilization of small-scale mountain farming, so prevalent in this region;
- The cultural heritage of the Røros area which – as a cultural capital asset – was converted into a business idea;
- The international trend towards green tourism and adventure tourism, which developed as the tourist business was maturing;
- The 'Protestant work ethic' and local patriotism of the mountain region, which may be seen as a competitive mental advantage; and
- Some remarkable personalities, both in the private and in the public sector, who from the 1960s onwards had the vision and force of character to *choose development as a political and economical strategy.*

Why Has the Island Region Experienced Steady Decline?

How then could we explain why the coastal region did not fare so well? In the 1960s and 1970s Hitra and Frøya did not do very well as far as economic development was concerned. Youth, and people employed in the primary industries, left the region and unemployment rates were high. At the time, both regions were 'problem regions'. However, unlike the Røros region, the coastal region could not find any common strategy to meet these challenges. At the same time as the combination of fishing and farming was breaking down because of the tendency to specialization both in farming and in the fisheries, the fish resources were being depleted. First, the important winter herring catch almost disappeared on the Western coast of Norway. Then, other species such as the cod and other white fish stocks were emptied because of over-fishing. As a result, the coastal fishing fleet struggled, and the only survivors were those who had enough capital to buy large trawlers, together with a few medium sized, family operated, boats. The family-organized fishing and farming combination disappeared altogether, which also undermined local civil society.

The first great positive change in Hitra-Frøya came with the new fish farming industry in the late 1970s. Actually, some of the first entrepreneurs in this whole industry were the Grøntvedt brothers from Hitra. Economically, the new salmon and rainbow trout industry was a great success, but as far as employment was concerned, it did not replace the number of jobs lost in the traditional fisheries. It also took 10-15 years before the industry was mature enough to compete on international markets. One major obstacle was transport costs: cooling lorries had to be taken on ferries to the mainland.

Tourism and other new services also experienced too low a growth rate to maintain the population. The concept of 'coastal tourism' developed very slowly, and it was only in the 1990s that this phenomenon has became known in the tourist markets as a generic label with distinct qualities. One reason for this is the traditionally low consciousness of the history and local heritage in the coastal communities, compared to that of the inland and mountain farming and mining communities.

The rather individualistic and cosmopolitan attitude in this coastal culture also gave the youth good reasons to leave and find their fortunes elsewhere. Co-operation between the neighbouring municipalities of Hitra, Aure and Snillfjord to build an oil industry land base at Tjeldbergodden in the late 1980s was a rather remarkable exception to the usual, individualistic, 'do it yourself' attitude. Perhaps this common effort, which has been very successful for all parties, has future potential. The new mainland road connection also holds possibility for co-operative efforts in transport and infrastructure. In the latter part of the 1990s, the coastal region seems to have been able to stem the depopulation tide.

Conclusions

If we consider the two cases together, seven factors appear to have contributed to the very different outcomes of the development process in these two regions:

- The concomitant breakdown of the family-based fishing and farming combination and the herring and cod resources at the traditional fish banks of Hitra and Frøya. As a consequence, traditional civil society on the islands experienced a strong decline, without any institutional innovation to replace it;
- Even though the copper resource was exhausted at Røros, the mountain region managed to find new niches in tourism, manufacturing industry and mountain farming;
- This break with the past and readjustment to the new terms of trade happened most of all because some remarkable leading political and administrative personalities acted on behalf of both the private and the public sector from the 1960s and onwards. They built organizations and

networks that were able to replace the institutional decline in primary and secondary sector;

- Hitra and Frøya are islands, which have excluded them from the most important production and service infrastructures in the last 30 years, namely roads and air traffic;
- The individualistic approach in the coastal communities, in contrast to the collectivist approach in the mountain region, has not been conducive to creating networks and common strategies to meet globalization's threats and possibilities;
- It seems that the coastal communities have had a consolidation period in the 1990s, which in many ways is what happened in the mountain area in the 1960s and 1970s. The mainland connection may function as a catalyst to further development; and
- Still we lack some evidence to conclude that the central actors of coastal region have chosen collaborative development as a strategy, as the main actors in the mountain region did in the 1960s and 1970s.

Notes

1 I am drawing evidence from data gathered in several projects carried out within these two regions by others (Gjerdåker, 1991; Daugstad *et al.*, 1999), as well as using data collected in my own research in the regions (Almås, 1977; 1980).
2 Central Bureau of Statistics, Population in Municipalities 1951-2000.

References

Almås, R. (1977), *Norsk Jordbruk – det nye Hamskiftet,* Gyldendal, Oslo.

Almås, R. (1978), *Hitra-aksjonen: Småbønder i Skattestreik,* Oktober Press, Oslo.

Almås, R. (1980), *Nye Organisasjonsformer i jordbruket,* University Press, Oslo.

Amin, A. (1994), 'The Difficult Transition from Informal Economy to Marshallian Industrial Districts', *Area,* vol. 26, pp. 13-24.

Amin, A. and N. Thrift (1994), 'Institutional Issues for the European Union: From Markets and Plans to Powers of Association', *Paper presented at the fourth General Conference of the European Science Foundation's RURE Programme,* Rome.

Asheim, B. (1991), *Flexible Specialisation, Industrial Districts and Small Firms: A Critical Appraisal,* in H. Ernste (ed.), *Regional Development and Regional Industrial Response: Extending Flexible Specialisation,* Belhaven Press, London.

Bourdieu, P. (1979), *La distinction. Critique Sociale du Jugemen,* Les Èditions de Minuit, Paris, English edition (1984) *Distinction: A Social Critique of the Judgement of Taste,* Routledge & Kegan Paul, London.

Bourdieu, P. (1993), *Sociology in Question,* Sage Publications, London.

Busch, L. (1994), 'Beyond Political Economy: Actor Networks in the Globalisation Of The Agro-Food System', *Presented at the International Conference on the Restructuring the Agro-Food System: Global Processes and National Responses*, Trondheim, 2-4. May.

Daugstad, K. *et al.*, (1999), *Bergverksbyens Omland. Om Ressursbruk, vern, Kultur og Natur i Rørosområdet*, NINA-NIKU, Trondheim.

Duncan, C. (ed.) (1992), *Rural Poverty in America*, Auburn House, New York.

Gjerdåker, B. (1991), *Kystsamfunn Skifter Ham. Med Hitter – Og Frøyværing Inn I Den Nye Tid*, Doctoral thesis, Norwegian University of Science and Technology, Trondheim.

Granovetter, M. (1973), 'The Strength of Weak Ties', *American Journal of Sociology*, vol. 78, pp. 1360-1380.

Granovetter, M. (1985), 'Economic Action and Social Structure: The Problem of Embeddedness', *American Journal of Sociology*, vol. 91, pp. 481-510.

Johannisson, B. (1994), 'Lokalt Företagsamhet för Globalt Bruk', I *Framtidens Regionala Politik Kräver Nya Grepp*, ERU-rapport 82, Fritzes, Stockholm.

Latour, B. (1987), *Science in Action: How to Follow Scientists and Engineers Through Society*, Harvard University Press, Cambridge, MA.

Marsden, T. (1992), 'Exploring a Rural Sociology for the Fordist Transition. Incorporating Social Relations into Economic Restructuring', *Sociologia Ruralis*, vol. 32, pp. 209-230.

Marsden, T., Murdoch, J., Lowe, P., Munton, R., and Flynn, A. (1993), *Constructing the Countryside*, University College London Press, London.

McMichael, P. (1996), *Development and Social Change – a Global Perspective*, Pine Forge Oaks, Thousand Oaks, California.

Putnam, R. (1994), *Making Democracy Work: Civic Tradition in Modern Italy*, Princeton University Press, Princeton.

Pyke, F., Beccatini, G., and Sengenberger, W. (eds)(1990), *Industrial Districts and Inter-Firm Cooperation*, ILO, Geneva.

SINTEF (1999), *Norges Muligheter for Verdiskaping Innen Havbruk*, Det Kongelige Norske Videnskabers Selskab, Trondheim.

Ward, N. and Almås, R. (1995), 'Explaining Change in the International Agro-Food System', in N. Ward, and R. Almås (eds), (1995), *The International Agro-Food System: Global Restructuring and Local Change*, Critical Perspectives on Rural Change Series, Vol. IV, Fulton Publishers, London.

Chapter 12

Local Consequences and Responses to Global Integration: The Role of the State in the Less Favoured Zones

Manuel Belo Moreira

Introduction

Globalization is probably one of the most controversial subjects of current political and social life, involving dimensions and means so diverse that a careful delimitation of the parameters of any approach is required.

The aim of this chapter is to explore the situation resulting from the pressure that global capital, through the world financial market, has placed on the nation state, thereby constraining its ability to fulfil certain regulatory functions that, until recently, have been part of its main prerogatives. One of the main characteristics of the globalization process has been to affect change in the role of the state, with it becoming less the regulator – and more the facilitator – of the needs of capital. The focus of the chapter is to assess the importance, especially to rural areas, of the abandoning of regulatory functions by the nation state or, what amounts to the same thing, the state's growing lack of power to enforce them.

In the first section I describe, very briefly, the distinctive lines of the globalization process, its evolution and consequences. In the second section I give particular emphasis to the changes in the role of the nation state in an increasingly globalized world. The third section investigates the economic consequences that have resulted from these changes on a global level, reserving for the fourth section an examination of the more specific consequences that have been experienced by the less favoured rural areas. The fifth section underlines the need for state intervention and the chapter ends with a summary of the main conclusions, and with reference to the double challenge for rural areas – the threat of progressive marginalization versus the dangers inherent in adaptive restructuring.

The Distinctive Lines of Globalization

There is general consensus that the ideological and political changes that gave rise to the processes of liberalization and deregulation constitute, along with the technological revolution in transport and information technologies, the necessary conditions for the process of globalization.

These features lie at the origin of the contraction of time and space and facilitate the 'intensification of social relations of global scale' that characterize globalization (Giddens, 1992, p.50). The emergence of a truly global communication network (as understood in its broadest sense) should also be emphasized. The 'culture industry' provides an excellent example of this trend; indeed, growing concentration in the media sectors has meant that ideological control has, as well, become increasingly concentrated.

In assessing quantitative changes brought about by globalization, it is also worth mentioning the substantial growth of international trade that has taken place in recent years and especially the increasing importance of *intra-firm movements* (Watts and Goodman, 1997). Furthermore, the increasing number of transnational corporations (TNCs) – which increased over sevenfold in the last 20 years – together with the growth of foreign direct investment and, above all, the explosion of international financial movements provides additional evidence of these changes (Petrella, 1996; Nunes, 1998).

Besides the development of highly sophisticated and efficient forms of *multi-spatial co-ordination,* even at a great distance, emphasis should also be placed on the adoption of new strategies by the TNCs, including the appearance of new methods of management, that have caused the emergence of the *network enterprise* (Dicken, 1992; Castells, 1997) and *global commodity chains* (Gereffi and Korzeniewicz, 1994). There has also been a proliferation of strategic alliances to finance the enormous cost of research and development.

Finally, it is important to stress the profound and radically new changes being registered in the international finance system (Bourguinat, 1992; Beaud, 1997; Eichengreen, 1998). These modifications were only made possible after financial deregulation and liberalization had been initiated, that is, once the political decisions based on the hegemony of the neo-liberal ideology had been taken (Chesnais, 1994; Korten, 1996).

With deregulation, transnational financial companies and institutions have formed an 'ever denser network of buying and selling, credit and loan, hedging and speculative operations, that transcends national frontiers and creates the basis for the organization of a truly global financial mega-market' (Bourguinat, 1992, p.20). The hyper-mobility of capital and the growing power of institutional investors, namely saving funds and pension funds, which Aglietta (1992, p.61) designates *strategic money,* constitute the essential features of this global financial market.

In contrast, the development of the global finance market has been characterized by the explosion of virtual money movements and businesses operating 24 hours a day using software which is able to automatically react to certain stimuli. This has called into existence what Drucker (1993) denotes as the *symbolic economy,* thereby creating a source of volatility and instability that has enormous repercussions for the 'real' economy. One only has to look at the onset of financial crises to confirm this: Wall Street in 1987, followed by Mexico in 1994/95, then the crisis that began at the end of 1997 in Asia, and which spread to Russia, and which has also been felt in Latin American countries (particularly Argentina in 2002).

Perhaps the most serious problem is that labelled by Strange (1998) as *casino capitalism,* namely a globalized financial market that is subjected to a speculative logic and dynamic, but for which no effective control mechanisms exist. This global financial market is extraordinarily sensitive and responsive to any sign of state intervention, thereby imposing on the nation state what Fitoussi (1997) has referred to as *financial tyranny.* This source of instability is creating a shift in the perceptions about the consequences of liberalization of financial flows. Some leading economists such as Bhagwati (1998) and Krugman (1997; 1998), as well as politicians are urging for the imposition of the controversial Tobin Tax on exchange operations (Tobin, 1999) as a means of regulating (and allowing the nation state to profit from) capital flows.

The above-mentioned aspects are at the core of the concept of globalization, a process defined as a 'stage beyond *internationalization* in which forms of capital extend beyond the nation-state but where the nation-state remains the significant unit of analysis' (Friedland, 1994, p.30). Such a characterization of the globalization process justifies the need to carefully assess what changes have occurred in the role of the nation-state.

Changes in the Role of the State

While the reality of globalization clearly requires a reformulation of the concept of state (see McMichael and Myhre, 1991; Bonanno *et al.,* 1994a; 1994b), this is not the aim of the present chapter, the main intention of which is to emphasize that the 'withdrawal of State intervention in some spheres (particularly social spheres) has been accompanied by the strengthening of other spheres' intervention' (Bonanno *et al.*, 1994a, p.22).

Essentially, global capital has gained new degrees of freedom because of the imposition of *financial tyranny,* which has contributed to the weakening of the nation state, even if there is a considerable variation of situations from one country to another and between different areas of economic activity (Friedland, 1994). However, it is less important to know whether the state is gaining or losing power. Rather, the key question is 'which interests dominate its control and to what ends'? (Bell and Lowe, 2000, p.287).

In the discussion below I use as my starting point the characterization of the state proposed by McMichael and Myhre (1991), in which the state functions as a facilitator of capital requirements. It should be stressed that if, in a global context, it is possible to speak of the state as facilitator, when we look at concrete situations it is necessary to take local specificity and discrete historical and economic contexts into account. Nevertheless, for those countries that are already relatively more dependent on transnational capital and/or have come under IMF 'supervision' and/or have agreed programs of structural adjustment with the IBRD, economic priorities often dictate a strengthening role for transnational capital. When 'stabilization and adjustment' programmes are adopted in those countries, the resulting depreciation of national currencies in relation to the dollar makes the purchase of national assets more attractive to outside investors (Amin, 2000).

For obvious reasons, especially in situations of high unemployment, the state plays the role of facilitator on a somewhat selective basis, welcoming those fractions of transnational capital that show the greatest willingness to establish new facilities and/or to take over and modernize industries with low or declining competitiveness. While we are used to seeing the state offering its 'facilities' to the highest bidder, under conditions of globalization, the situation is reversed – with transnationals auctioning – on an international basis – their availability and investment capacity, in an effort to determine which government might offer the best conditions for global capital.

In contrast to this, the state dare not risk regulating speculative financial capital (usually the province of the financial divisions of large transnationals)(see Chesnais, 2000). In fact, it is important to acknowledge that financial liberalization, once adopted, is largely out of the control of the state. Very few states that are well integrated into the world market are able to escape the constraints imposed by the power of financial capital, although significant differences between countries can be found.

A state's agreement to facilitate global capital requirements is obviously not made on the basis of equal international power relations, nor are the results neutral. This is the case, whether the state adopts the role of its own volition, or has it directly or indirectly imposed by hegemonic states and/or institutional mechanisms. For example, IMF conditionality means that financial assistance is offered in return for privatization, internal market liberalization and the opening up of the economy to greater import and investment penetration. The total satisfaction of transnational capital requirements has two main consequences.

The first is the loss of control over financial movements, which has implications for the way(s) in which crises may be addressed, as well as something that economists usually fail to assess, namely the diminishing control over the laundering of money and the growing potential for the evasion of taxes. Losing control over financial movements is equivalent to submitting to the 'casino economy' since the capacity of the nation state to deal with crises is weakened and it becomes impotent in the face of financial crime (Zey, 1993; US State Department, 2000). The 1990s Asian crisis shows that in countries such as China and India, where restrictions are still imposed on capital movements, the impact was less marked than in other cases.

The second is the obligation to accept financial orthodoxy, so esteemed by the markets. This is equivalent to (a) foregoing the regulation of economic activities within national space, (b) accepting the maintenance of balanced budgets and balanced external accounts, (c) cutting social (particularly welfare state) expenditures, (d) reducing taxes and (e) promoting privatization.

Even those nation-states with a more interventionist tradition have lost much of their capacity to determine the trade-off between welfare and employment. This means that the distribution of incomes shifts, with the share of capital increasing relative to that of labour, reversing a long-run tendency that had characterized the richer European countries before this phase of globalization began. The need to comply with financial tyranny is pressuring the more interventionist European states to abandon the continental social democratic model and adopt a more liberal form of capitalism.

Furthermore, the creation of the Euro necessitated the establishment of a European Central Bank with a disproportionate priority given to purely financial concerns. This has meant that a quite specific institutional form and operational practice have emerged, in which European policy makers have relinquished all significant tools of macroeconomic management. Once again, the neo-liberal hegemony has prevailed, and the adopted model is even more restrictive than the US model. From now on the EU will not be able to put in practice any significant Keynesian policies generalized to all its members, unless the Commission were to alter the mandate of the European Bank – an initiative which could itself provoke financial turmoil.

The Consequences of the Diminishing Extent of State Autonomy

The consequences of a reduction in the extent of state autonomy caused by globalization differ quite markedly depending upon the country or socio-economic context in question. This distinction becomes clear if we consider the polar opposites of the continuum between rich, industrialized countries with a strong tradition of state intervention, on the one hand and the poor, non-industrialized countries, in which the few existing institutions are weak, inefficient and ineffective, on the other. It is one thing to envisage the possible outcomes of a reduction in state intervention in the first case, where the question is essentially one of a change in the intensity and/or extent of intervention, or in the traditional form of intervention. It is quite another to imagine the impact of globalization on the state's role in countries where the classic state institutions do not function with minimal standards of efficiency or, indeed, in the extreme case, in which the majority of the institutions that characterize a sovereign state barely exist at all.

In general terms – and from the perspective of global capital – the institutions of some interest are those designed to ensure the fullest possible functioning of the market. Yet, amid the neo-liberal exhortations that the state progressively withdraw from all forms of intervention, free market adherents tend to forget that without the state there is no market! This issue has been explored by Mingione, (1991) following the lead of Polanyi. It is also being explored by economists like Stiglitz (1998a; 1998b) and Rodrik (2000). The latter offers a revealing case of the importance of public institutions in facilitating market function. In addition to institutions defining and defending property rights, he also refers to regulatory, macro-economic stabilization, social insurance and conflict management institutions.

Foreign investors' investment decisions increasingly depend on how attractive (or otherwise) they find the overall cost structure in particular countries or territories. Since everything that hampers the efficient functioning of the above-mentioned institutional arrangements causes a rise in transaction costs, their existence and functioning is of crucial importance both to inward investors and to attempts by the state to attract foreign capital. In fact, provided that minimal conditions are guaranteed, global capital scrutinises further conditions on which the decision to invest, disinvest, or relocate depends. Such criteria are systematically applied by TNCs and provide continuous pressure on national governments to offer additional stimuli to inward investment, such as fiscal exemptions, provision of dedicated infrastructure, and so forth. In short, the expectation of global capital is that the state will guarantee the regular functioning of the market economy, while maintaining the weak profile of facilitator.

However, from the standpoint of the state, and under such circumstances, in order to make market-sustaining public institutions work effectively, what is required is an appropriately-resourced and well-organized public administration, with competent staff and the political will to enforce regulations. While this pre-condition is easy to formulate in the abstract, it is much more difficult to put into practice, especially when the above-mentioned constraints on state intervention are taken into account.

It is important to emphasize, too, that globalization tends to aggravate what O'Connor (1973) called the fiscal crisis of the state. This is mainly due to the reduction in taxation of profits and capital gains (see Chesnais, 2000) providing the state with limited funds to invest in public services, upon which state legitimacy rests. Part of the problem is that process of reducing expenditure can be arbitrary and its effects unpredictable, unless public administration is very efficient.

The overall consequence of the above constraints is that the state has fewer resources to build up and maintain market-sustaining public institutions. The problems, therefore, are the following: how does the state correct the externalities provoked by the market failures, and how does it create, deliver and guarantee the public goods necessary for the functioning of the economy?

Correcting negative externalities means forcing economic agents to internalize them: correcting positive externalities means compensating economic agents for the costs incurred in producing them. Since many of the consequences of globalized capital cannot be confined within national borders, it is beyond the power of individual nation states to correct global externalities and create global public goods. Unfortunately, the question of what kind of institutions may be able to fulfil this function has yet to be seriously addressed.

It must be remembered that while huge and powerful TNCs may not need the same level of assistance as small firms or infant industries, this does not mean that they do not actively seek it. However, the 'bottom line' is that they still need state regulations, as well as the capacity to enforce them. This conclusion underpins the strategy that global capital has adopted, namely:

- Nation-states are put under pressure to withdraw from regulatory tasks, including environmental controls and, especially, financial supervision, and to accept reductions in state expenditure;
- A the same time, however, global capital needs to assure a certain level of regulation, and needs to have guarantees that the state will enforce it;
- Clearly, the above requirements are difficult to fulfil, since the effects of a squeeze on expenditure are usually detrimental to the state's regulatory means and capacity;
- So, with a view to guaranteeing that governments fully comply with its requirements, the strategy of global capital is to use the global regulations of the WTO (World Trade Organization) and/or the proposed MIA (Multilateral Investment Agreement) to hold governments and populations 'hostages'.

This perception of the arrogant behaviour of global capital fuelled the anti-globalization feelings expressed by different sections of civil society at various WTO and other 'summits' of global regulators. Yet, national governments face significant difficulties in opposing global forces. Decision-makers for global capital know that if the nation states do not comply with its needs, as might happen were a new government is elected with an anti-globalization programme, then it would be required to pay such indemnities as the global regulations specify. It must do this in order to avoid both a significant reduction in its global credit rating, and the attribution of financial pariah status.

The conclusion is clear. On the one hand, globalization is contributing significantly to a weakening of the ability of the nation state system to correct local externalities. On the other hand, since the reach of the market is every day more global, most externalities are global in character, thereby creating a situation that no single nation state can properly address. In fact, while, at the national level, it is already very difficult to create and maintain the necessary public goods and to create the mechanisms to correct externalities, it is even more difficult to do so at the global level. As a consequence, an enormous effort in negotiation is needed to create a viable basis for the market-sustaining institutions necessary both to provide and deliver global public goods and to correct global externalities (Kaul *et al.*, 1999).

Consequences of Global Integration for the Less Favoured Rural Zones

In order to evaluate the consequences of global integration on the less favoured rural zones, the logic and dynamic of the economic agents needs to be considered according to the following rationale:

- Local economic activities in these zones suffer from growing competition, not only from traditional local competitors and the regulation-evading activities (informal economy), but from enterprises competing from further afield and penetrating the new localities. This means that the typical small-scale forms of production in these zones (small capitalist firms, handicrafts, family farms and co-operatives) have to face cut-throat competition from forms of production that are quite distinct, some of them integrated into global commodity chains. However, for small-scale local businesses themselves access to these global chains is extremely difficult: only in exceptional circumstances are they are integrated into such global networks; and
- Globalization has accentuated the trend towards the concentration of capital, mainly due to economies of scale and scope. This, in turn, encourages the geographical concentration of production and a consequent abandonment of the less favoured rural zones by businesses with strong scale economies.

A number of examples from the complex of agribusiness *filières* will serve to illustrate the logic and dynamic of the process of economic marginalization suffered by less favoured rural areas.

The production of fruit and fresh vegetables is one of the sectors where globalization has the most visible effects, to the point that the concept of the fruit season is, nowadays, meaningless. Thus, certain regions – producing varieties of fruit that previously enjoyed a secure market because of their ability to meet demand at 'irregular' times of the year – have had to deal with competition from out-of-season fruit, imported without limits, through global chains endowed with massive economic and financial power. If the advertising power of these global chains is also considered, it is easy to understand why many varieties of agricultural products have begun to come under pressure in what were formerly their traditional markets. This, at the extreme, could mean that these products will eventually stop being produced in the less favoured rural zones, unless those interested can find efficient ways of responding to global competition.

A further example is that of the cereals and animal feed sector. Powerful TNCs, connected through vertical integration processes with subcontracting meat producers, mainly pork and poultry (see Hefferman *et al.*, 1999) are involved in intensive production. They are intimately connected to the animal feed sector, and the more extensive ranch system that is, nevertheless, also experiencing huge capital concentration. The impact of such agro-food complexes on the small-scale production in less favoured areas has tended to be the same as that of the fruit and fresh vegetables sector.

In the dairy sector, the role of larger TNCs, such as Nestlé, BSN-Danone, as well as Parmalat and Yoplait, are strengthening their weight in the kind of dairy production that promise higher margins, such as the production of yoghurts, dairy desserts and dietary products. Those lines with lower margins, and the direct contact with milk producers, are activities left to smaller private units or co-operatives. It is also important to acknowledge the strength of large TNCs in the beverage sector, with such well-known brand names as Coca-Cola, Pepsi, Nestlé, Perrier, and Seagrams increasing market share over time.

In short, in this context of globalization, highly localized economic activities (production, processing or marketing) are suffering from the entry of new competitors and/or the reinforcement of traditional competitors that are more closely connected to global commodity chains. Competition is so cutthroat that many small firms will be driven out of the market. In this sense, globalization has not started anything new, but it is certainly reinforcing and quickening the pace of a process of capitalist restructuring that started some time ago.

Thus, from the perspective of the less favoured rural zones, globalization obliges local forms of production to contend with the following dilemma: either to seek integration into a global commodity chain, which is difficult, and can be very dangerous since it strongly diminishes their autonomy and room for manoeuvre; or to find new ways of operating in the market. If neither occurs, economic marginalization will inevitably intensify, with population decline and social exclusion quickly following on its heels.

In short, globalization is a powerful device to force agricultural restructuring upon less favoured rural areas – something which has important spatial implications. Nevertheless, it must be stressed that such processes, in real life, are never so schematic and linear. In fact, the outcomes of the process depend significantly on the particularities of the social, historical, and economic context, as well as on the levels and efficiency of state intervention in the specific case under scrutiny.

For the purposes of this argument (and if we maintain the more schematic approach outlined above) it seems that – in broad terms – two distinct potential paths can be observed. One is the path of increasing marginalization for the rural economy resulting from the unfettered operation of market forces; the other is the path that call be described as positive restructuring.

Positive Restructuring

Theoretical considerations and empirical evidence both point to the need for a mixture of conditions to achieve positive restructuring. One thing is also clear: agriculture can no longer be the sole solution to problems of economic and social marginalization – at least in the European context (Baptista, 1993; Arnalte *et al.*, 1998).

The most successful cases of positive restructuring have involved the transformation of old ways of producing and marketing traditional commodities, as well as the introduction of new activities, including both production and services. Successful adaptation to the new era of globalization has demonstrated that:

- Change agents must be able to identify new processes of production and/or marketing, that is, new niche markets or ways of being more successfully integrated into global commodity chains; and
- The surrounding economic context has to be favourable to existing, as well as new, products and services that are offered by these agents.

To be successful means to take advantage of the new windows of opportunity opened up by globalization. It usually means exploring a number of features closely linked to traditional production, be it food, drink and cuisine, access to local culture, including the historical built environment and natural amenities. The question remaining, however, is who will pay for the existing traditional products/services, and/or the new ones?

The answer to this question demands that certain macroeconomic considerations be taken into account. The rationale of positive restructuring that I am describing, shows that it will be immeasurably easier to achieve this goal in a milieu that is relatively rich rather than one that is relatively poor. In fact, less favoured rural zones in the wealthier countries have easier access to a greater number of urban residents eager to have the opportunity of enjoying rural life, consuming rural products and benefiting from rural environmental amenities. Furthermore, these urban strata have enough purchasing power to enable considerable expenditure on such items.

But what happens in the less favoured rural zones located in the poorest countries? These constitute a completely different context in which the above-mentioned ways of achieving positive restructuring might occur. In fact, in most cases, there is little opportunity for these regions to advance via globalization. In these cases, agriculture is probably the sole economic activity in rural society and niche markets are almost non-existent and very difficult to reach. Some do manage to integrate their production successfully into global commodity chains, but many are left with survival strategies that tend to be increasingly disconnected from the global system.

The Necessity of State Intervention for the Positive Restructuring of Less Favoured Rural Zones

There is some doubt that market forces, if left alone, will be able to initiate positive restructuring. If marginalization trends can been reversed and success stories identified, analysed and promoted, external capital may be attracted. However, capital is unlikely to risk investment so long as the trend is still towards marginalization.

In order to initiate positive restructuring a number of necessary conditions must be met: funds must be provided for investment; there must be investment in the knowledge economy; and, above all, state intervention is essential if negative externalities are to corrected, and the necessary suite of public goods is to be established and sustained.

Concerning the financing of positive restructuring, it should be remembered that globalization has also increased the competition for capital and, nowadays, there are much more attractive alternative uses to which private funds can be put. Furthermore, commercial banks are not particularly interested in small businesses, especially those located in marginal zones (Desai and Mellor, 1993). So, from the point of view of the funding required, the need for state intervention is fully justified, even if only as a guarantee. An example here is the case of the US *Small Business Administration* programme, in which up to 95 per cent of the loans are guaranteed in the case of bankruptcy (Reifner and Evers, 1995).

Positive restructuring needs also to facilitate access to new knowledge and to train skilled labour. Yet, it is clear that in areas subject to marginalization there is a lack of human and social capital. It is often the case that people who are capable of identifying bottlenecks and restructuring traditional production and marketing are simply absent from the scene. There is a need to encourage new talent – people to create new products or services in response to the more demanding consumption patterns of incomers migrating from urban to rural areas.

It is clearly utopian to wait for market forces to initiate positive restructuring in order to reverse the very marginalization that those same market forces have helped to bring about! Institutions are needed to correct market failures and to provide public goods and to promote positive restructuring. The challenge is to find the kind of institution that, while being promoted and supported by the state, will avoid the failures and frailties that have so often been synonymous with government intervention.

In conclusion, the involvement of the state becomes daily more necessary, not only to correct market failures, but also to correct the lack of interest by the capitalist marketplace in the less favoured rural zones.

Conclusion

The great quantitative and qualitative transformations that are linked to globalization point to significant changes within the nation state. In this chapter the focus has been on the economic problems of less favoured rural zones, and the role of the state in 'correcting' the problem of marginalization.

The interpretation of the consequences of globalization focused on trying to understand the logic and dynamic of global capital, together with the impacts of global finances on the extent of autonomy of state intervention.

A number of difficulties that the state faces due to the globalization process were highlighted. First, the state faces increasing problems in trying to correct externalities. It faces difficulties providing an environment in which firms responsible for negative externalities are forced to bear the cost, and in which firms generating positive externalities are supported and encouraged. Second, the state also finds it increasingly hard to create and deliver the public goods and services the majority of people need – a difficulty that is even greater when one takes into account the limited human and social capital that remain in the less favoured zones.

Attention was also directed to the fact that the logic and dynamic of marginalization is the result of the market forces, and that globalization, with its trend toward increasing concentration of capital has substantially aggravated this process. Only two alternative outcomes can be predicted: continuing marginalization of the rural areas, or their positive restructuring.

Given that market forces are not interested in initiating positive restructuring in such localities, the need for state intervention becomes paramount. That is, in terms of less favoured rural zones, the state's intervention is not only necessary in order to correct market failures, but also essential to compensate for the lack of interest that global capital shows in the future economic and social viability of these areas.

Acknowledgement

I am in debt to Chris Gerry for his comments on a previous version of this text.

References

Aglietta, M. (1992), 'Libération Financière, Stabilité, Croissance', *La Nouvelle Revue Socialiste,* Juin, pp. 59-68.

Amin, S. (2000), 'Economic Globalism and Political Universalism; Conflicting Issues?' *Journal of World-Systems Research,* vol. 3, Fall/Winter, pp. 581-622.

Arnalte, E., Baptista, F., Lourenço, F., Muñoz-Zamora, C. and Rodrigo, I. (1998), *El Desarròllo Rural: Políticas Aplicadas, Situación Actual del Debate y Perspectivas del Futuro a Nivel Europeo,* Parlamento Europeo, Documento de Trabajo. AGRI-106.

Baptista, F. (1993), *Agricultura, Espaço e Sociedade Rural,* Fora de Texto, Coimbra.

Beaud, M. (1997), *Le Basculement du Monde,* Éditions La Découverte, Paris.

Bell, M. and Lowe, P. (2000). 'Regulated Freedoms: The Market and the State, Agriculture and the Environment', *Journal of Rural Studies,* vol. 16(3), pp. 285-294.

Bhagwati, J. (1998), 'A Liberalização é um mito', *Economia Pura,* Lisboa, vol. 1(3), pp. 84-87.

Bonnano, A., Friedland, W., Llambí, L., Marsden, T., Moreira, M. and Schaeffer, R. (1994a), 'Global Post-Fordism and Concepts of the State', *International Journal of Sociology of Agriculture and Food,* vol. IV, pp. 11-29.

Bonnano, A., Busch, L., Friedland, W., Gouveia, L. and Mingione, E. (eds), (1994 b), *From Columbus to ConAgra. The Globalization of Agriculture and Food,* University Press of Kansas, Lawrence, pp. 1-26.

Bourguinat, H. (1992), *Finance Internationale,* Presses Universitaires de France, Paris.

Castells, M. (1997), *The Rise of the Network Society,* Blackwell, Oxford.

Chesnais, F. (1994), *La Mondialisation du Capital,* Syros, Paris.

Chesnais, F. (2000), *Tobin or not Tobin? Uma Taxa Intrenacional Sobre o Capital Para Apoio aos Cidadãos,* Campo da Comunicação, Porto.

Desai, B. and Mellor, J. (1993), *Institutional Finance for Agricultural Development. An Analytical Survey of Critical Issues,* International Food Policy Research Institute, Washington, D.C.

Dicken, P. (1992), *Global Shift: the Internationalisation of Economic Activity,* 2nd edition, Paul Chapman Publishing Ltd, London.

Drucker, P. (1993), *As Fronteiras da Gestão,* 2ª. edição, Editorial Presença, Lisboa.

Eichengreen, B. (1998), *Globalizing Capital. A History of the International Monetary System,* Princeton University Press, Princeton, New Jersey.

Fitoussi, J-P. (1997), *O Debate-Tabu. Moeda,* Europa, Pobreza, Lisboa, Terramar.

Friedland, W. (1994), 'Globalization, the State, and the Labor Process', *International Journal of Sociology of Agriculture and Food*, vol. IV, pp. 30-46.

Gereffi, G. and Korzeniewicz, M. (eds), (1994), *Commodity Chains and Global Capitalism*, Wedtport, Greenwood Press, Connecticut.

Giddens, A. (1992), *As Consequências da Modernidade*, Celta Editora, Oeiras.

Heffernan, W., Hendrickson, M. and Gronski, R. (1999), *Consolidation in the Food and Agriculture System*, Report to the National Farmers Union, US.

Kaul, I., Grunberg, I. and Stern, M. (eds), (1999), *Global Public Goods. International Cooperation in the 21ˢᵗ Century*, UNDP and Oxford University Press, New York.

Korten, D. (1996), *Os Fracassos de Bretton Woods*, in Mander e Goldsmith (ed.), *Economia Global Economia Local*, Instituto Piaget, Lisboa, pp. 27-37.

Krugman, P. (1997), 'Seven Habits of Highly Defective Investors', *Fortune,* 29 December, pp. 22-23.

Krugman, P. (1998), 'Saving Asia: It's time to get RADICAL', in *Fortune,* 7 September, pp.27-32.

McMichael, P. and Myhre, D. (1991), 'Global Regulation vs. the Nation-State: Agro-Food Systems and the New Politics of Capital', *Capital and Class,* vol. 43, Spring, pp. 83-105.

Mingione, E. (1991), *Fragmented Societies: A Sociology of Economic Life Beyond the Market Paradigm*, Basil Blackwell, Oxford.

Nunes, R. (1998), 'Activos Estrangeiros Procuram-se', in *Economia Pura*, Dezembro, pp. 37-41.

O'Connor, J. (1973), *The Fiscal Crisis of the State*, St. Martin's Press, New York.

Petrella, R. (1996), 'Globalization and Internationalisation. The Dynamics of the Emerging World Order', in Boyer e Drache (ed.) *States Against Markets*, Routledge, Londres, pp. 62-83.

Reifner, U. and Evers, J. (1995), 'Small Business Lending in Regional Development – Northern Approaches and Programs in Development Financing', Comunicação à *International Conference Credit and New Entrepeneurs*, OCDE – LEED Program, IRED e Società per l'Imprenditorialità Giovanile S.p.a., Roma, 30 de Junho – 2 de Julho.

Rodrik, D. (2000), 'Development Strategies for the Next Century', Paper presented at the Conference on *Developing Economies in the 21ˢᵗ Century*, Institute for Developing Economies, Japan External Trade Organization, 26-27 January, in Chiba, Japan.

Stiglitz, J. (1998a), 'Distribuição, Eficiência e Voz: Elaborando a Segunda Geração de Reformas', in: http://www.dataterra.org.br/Documentos/ stiglitz2.htm.

Stiglitz, J. (1998b), *Em busca de um Novo Paradigma para o Desenvolvimento: Estratégias, Políticas e Processos*, in: http://www.dataterra.org.br/Documentos/ stiglitz_port.htm.

Strange, S. (1998), *Mad Money*, University Press, Manchester.

Tobin, J. (1999), 'Os Economistas Melhoraram a Condição da Humanidade, entrevista ao Courier da Unesco', *Economia Pura*, Ano I, nº 12, Abril.

US State Department, (2000), 'Money Laundering and Financial Crimes', in http://www.state.gov/g/inl/rls/nrcrpt/2000/index.cfm?docid=959.

Watts, M. and Goodman, D. (1997) 'Agrarian Questions. Global Appetite, Local Metabolism: Nature, Culture, and Industry in Fin-De-Siècle Agro-Food Systems', in D. Goodman and M. Watts (eds) *Globalising Food*, Routledge, London, pp. 1-32.

Zey, M. (1993), *Banking on Fraud: Drexel, Junk Bonds, and Buyouts*, Aldine de Gruyter, New York.

Chapter 13

Global Change, Social Exclusion and Sustainable Livelihoods in Rural Britain

Mark Shucksmith

Introduction

This chapter discusses the ways in which economic, social and political forces for change operate in rural areas to produce social exclusion for some people and social groups. After a brief review of these forces of change in Europe, the meaning of the term 'social exclusion' is considered. Following this, the chapter draws upon recent empirical studies to illustrate how social exclusion occurs in rural areas of Britain. The chapter draws attention to the effects on sustainable livelihoods of the ascendancy of market processes, and the waning of state systems, as a result of the neoliberal hegemony which has hastened deregulation, privatization, reductions in public expenditure and the strengthening of global capital's penetration of labour and product markets.

Forces for Change in Rural Areas of Europe

The economies and societies of rural areas of Europe are changing rapidly in the face of globalization, economic restructuring, migration, and other social and policy changes. These forces have different implications for different areas and different social groups, in a wide diversity of rural contexts.

Market and Economic Forces

Many rural areas (such as those in Scotland) are now growing faster than urban districts, while others experience decline. Because there is no standard trend, the economic and social processes are not always well understood. One key element, however, is the increasingly global penetration of local markets. On the one hand, international capital may seek to exploit those rural areas characterized by low wages, a compliant, non-unionized workforce, and lower levels of regulation, leading to increased dependency and peripherality. On the other hand, local capital may seek to develop products which depend upon a local identity for their market niche, so 'selling the local to the global'.

According to the European Commission (1997, p.15), 'agriculture and forestry no longer form the backbone of rural economies throughout the EU.' Agriculture still employed 16 million people in Europe in 1993, but this constituted only 5.5 per cent of total EU employment, and even in the most rural regions its share in 1990-91 was only 12 per cent. The declining importance of agriculture and other primary activities has been more than offset in many rural areas by the growth of services. Thus, the EC (1997, p.16) highlights some rural areas as the most dynamic in the EU. Around 73 per cent of jobs in rural Britain are now in services – compared to 60 per cent in 1981 – notably in public services, education, health, distribution, tourism, and financial services. Rural areas have shared in a general shift to a service-based economy in which the information and knowledge-based industries play an increasing role, bringing both opportunities and threats.

The EC (1997, p.16) concludes that 'rurality is not itself an obstacle to job creation: ...it is not synonymous with decline'. Most rural areas in the UK, for example, have coped well with the need for change. 'Employment in rural areas has increased more rapidly than in other areas...[and] unemployment in rural areas is generally lower' (Cabinet Office, 2000). This may be misleading, however, in so far as Beatty and Fothergill (1997) show that unemployment is systematically under-reported in rural Britain, and this is likely to be the case elsewhere. Moreover, some areas have found it harder to adjust to rapid restructuring, notably those which are remote and have a high dependence on agriculture or other primary activity. Even where new jobs have appeared, some people have found it hard to adjust. A particular feature of rural employment in Europe is the prevalence of small firms. Over 90 per cent of all rural firms in Britain, for example, are micro-businesses, employing fewer than 10 people, and 99 per cent employ fewer than 50.

The European Commission (1997, p.14) notes that, 'over the coming years, the capacity of rural areas to maintain or create jobs will have a major impact on the unemployment rate and/or migration flows'. Given that these jobs are not likely to be in agriculture, that report goes on to suggest (1997, p.16) that 'the creation of rural employment results from a specifically territorial dynamic which may not yet have been systematically analysed at EU level, but which seems to include such features as:

- A sense of regional identity and social cohesion;
- An entrepreneurial climate, a capacity to link up with the economic mainstream, public and private networks;
- A good educational level; and
- An attractive cultural and natural environment'.

These may be summarized as cultural, social, human and natural capital. Their role in rural development is discussed further in Shucksmith (2000a).

Demographic and Social Changes

Fundamental demographic, social and cultural changes also characterize rural areas in Europe. Migration flows are critical and, while some areas continue to lose population, in many parts people are moving into rural areas because of the new values placed on rural space (including a clean environment, healthy lifestyles, and strong community interaction). The consequences of the imposition of such values on rural societies may be far reaching. Across the EU, 46 per cent of predominantly rural regions are growing, while 42 per cent are in decline; and of significantly rural regions 57 per cent are growing, while only 34 per cent are declining (EC, 1997, p.10).

This migration tends to be socially selective. Gentrification has been evident in many accessible or tourist areas of rural Europe, as the affluent middle classes have migrated to the countryside, perhaps displacing less affluent groups (cf. Phillips, 1993, for evidence of this process in Britain) through competition for scarce housing. Much has been written about the rise of a rural professional and managerial 'service class' such that certain regions may be colonized by knowledge-workers at a distance from production activities. Even in some attractive remoter areas, retirement migration and distance-working (working from home) may produce similar effects, though in less attractive (or ex-industrial) rural areas, with low wages and low rents, low-grade jobs may be all that can be attracted. Migration also tends to be age-specific, with young people often leaving rural areas, as discussed later in this chapter, and older families moving in.

Social relations are also changing in other ways with the rise of individualist values and the decline of established institutions. Higher divorce rates, delays in the age at which people get married and have children, and increasing life expectancies all tend to lead to a decline in the average size of households and to a greater demand for houses. Moreover, changes in the age structure of the rural population, together with the economic restructuring described above, are tending towards increased dependency ratios, casualization, part-time working, and less job security. The interactions between these changes, within the family and in employment, are not well understood in rural contexts.

Changing Policy Contexts

Rural policies are changing in response to these forces, and many wider policies (especially macroeconomic policies and social policies) also have pervasive impacts upon rural areas. For example, many countries in Europe recently experienced protests about the impact of high fuel taxes in heightening social exclusion in rural areas.

Social policy and welfare reforms are particularly important in addressing inequalities and in offering support and opportunities to the most disadvantaged. In Britain, the Labour government's welfare reforms have provided both incentives and pathways towards labour market integration, facilitated by the expansion of the economy and an associated increase in the aggregate demand for labour. Its 'New Deal' has sought to address, in the first instance, the integration of young people into work and this has faced particular obstacles and challenges in rural areas, notably arising from the small size of rural firms, the distances involved, and the low levels of skills required. There are also challenges in delivering personal counselling (the gateway to the New Deal) in some rural areas. The New Deal is now being extended to several other groups including lone parents, and those of working age over 50. Similar national policies operate in most countries of the EU, following the requirement for National Action Plans to address unemployment, as agreed at the Luxembourg summit.

European policies are particularly important in relation to agriculture. Farmers receive large subsidies, and agricultural spending dominates the EU's expenditure. A reform of the EU's Common Agricultural Policy (CAP) has recently been agreed and this will reduce price support, tariffs on imports and export subsidies while partially compensating farmers for these through enhanced direct payments. Increasingly, these will become linked to environmentally sensitive farming and to areas facing particular hardship. More fundamental reforms appear inevitable, in the context of EU enlargement and World Trade Organization (WTO) negotiations, with declining support to farmers unless linked with rural development or environmental goals.

During the 1990s there has also been a tendency across Europe towards an increasing emphasis on capacity-building and community development in rural policy, informed by the EU's LEADER pilot initiative on rural development.[1] It is claimed, officially, that such an approach will permit innovative solutions to be developed for rural problems, by combining three elements: a territorial basis; the use of local resources; and local contextualization through active public participation. Endogenous development of this form is seen as building the capacity of localities or territories (though not necessarily of all individuals) to resist broader forces of global competition, fiscal crisis or social exclusion. To some extent, this similarity of approach to rural development may reflect a Europeanization of member states' rural policies (Shortall and Shucksmith, 1998).

But how do these forces for change connect with the individual experiences of people living in rural areas? To answer this, we turn to the concept of social exclusion, best understood, according to Byrne (1999) through an investigation of the intersection of history and biography.

Conceptualising Social Exclusion

The concept of social exclusion is contested, and no single agreed definition exists. The term has been used in three principal senses (Levitas, 1998):

- As an **'integrationist'** approach in which employment is seen as the key integrating force, both through earned income, identity and sense of self-worth, and networks;

- As a **'poverty'** approach in which the causes of exclusion are related to low income and a lack of material resources; and

- As an **'underclass'** approach in which the excluded are viewed as deviants from the moral and cultural norms of society, exhibit a 'culture of poverty' or a 'dependency culture' and are blamed for their own poverty and its intergenerational transmission.

This chapter takes an amended integrationist approach in the belief that this is best suited to developing an understanding of processes of exclusion, but that these extend far beyond labour markets and indeed are multi-dimensional (Shucksmith and Chapman, 1998).

Poverty is usually viewed as an outcome, denoting an inability to share in the everyday lifestyles of the majority because of a lack of resources (often taken to be disposable income). In contrast, social exclusion is see.

as a multi-dimensional, dynamic process which refers to the breakdown or malfunctioning of the major systems in society that should guarantee the social integration of the individual or household (Berghman, 1995). It implies a focus less on 'victims' and more upon the processes that cause exclusion. It also acknowledges the importance of the local context in such processes. Thus, while the notion of poverty is *distributional*, the concept of social exclusion is *relational*.

A particularly fruitful way of viewing processes of social exclusion and inclusion is as overlapping spheres of integration.[2] In a similar approach to Kesteloot (1998), Duffy (1995) and Meert (1999), Reimer (1998) argues that it is helpful to distinguish the dimensions of social exclusion according to the different means through which resources are allocated in society. He proposes four systems, however, which better capture the different processes that operate. They are as follows:

- Private systems, representing market processes;
- State systems, incorporating authority structures with bureaucratic and legal processes;
- Voluntary systems, encompassing collective action processes; and
- Family and friends networks, a system associated with reciprocal processes.

One's sense of belonging in society, as well as one's purchase on resources, depends on all these systems. Indeed, some have argued that these form the basis of citizenship.

This analysis recalls Polanyi's (1944) work on household survival strategies in relation to three spheres of economic integration: market exchange; redistribution or 'associative relations'; and reciprocity. Polanyi argued that the main form of transaction other than the market is reciprocity based on mutual affection and love, most notably within the family or household.[3] Reimer also relates his suggested four systems to the work of Fiske (1991), who proposed four 'elementary forms of human relation', namely market pricing, authority ranking, equality matching and communal sharing. Such a conceptualization of social exclusion in terms of the means by which resources and status are allocated in society requires, in turn, an analysis of the exercise of power.

Early research into disadvantage in rural Scotland (Shucksmith *et al.*, 1994; 1996), together with the Cloke *et al.*,'s (1994) rural lifestyles studies in England and Wales, identified processes of exclusion operating differentially in many rural areas of Britain. Labour markets and housing markets were instrumental in generating inequality and exclusion, with many respondents perceiving very restricted opportunities for well-paid, secure employment or for affordable housing, while at the same time these markets enabled affluent households to move into rural areas, drawing income from elsewhere. Young people and women tended to have the fewest options. These impediments to inclusion were closely bound up with failings of private and public services, most notably transport, social housing and childcare. Moreover, the welfare state was patently failing to reach potential recipients and the take-up of benefit entitlements was low. Access to advice and information in distant urban centres was problematic, and respondents were often confused about the benefits available and their entitlement. To mitigate these failings of markets and state, there was a greater reliance on the voluntary sector (which was itself under pressure as volunteers – mainly women – declined in number) and on friends and family. Migration and the loss of young people – also related to housing and labour market processes – ruptured informal support networks and left elderly people socially isolated. (See Philip and Shucksmith ,1999).

The very processes, then, which have supported the economic restructuring and gentrification of many rural areas, allowing rural areas to 'share in the nation's prosperity', have also created social exclusion and inequality. The way in which social exclusion has been conceptualized in this section holds out the hope of being able to connect the macro-level forces which operate to structure disadvantage and inequality with the micro-level experience of individuals in rural areas – that is, of being able to relate history to biography. The remaining sections of this chapter examine in more detail a number of arenas (income, employment, and access to housing) to illustrate the operation, and interaction, of these systems of inclusion and exclusion. These draw on research recently been funded and published by the Joseph Rowntree Foundation (JRF), for whom this author was programme director (Shucksmith, 2000b).

Incomes in Rural Britain: Poverty Amongst Affluence

Most previous research into rural poverty has emphasized counting the numbers of poor or disadvantaged people at a point in time. Yet, it is not enough to count the numbers and describe 'the socially excluded'. It is also necessary to understand and monitor the *processes* of social exclusion and to identify the factors that can trigger entry or exit from situations of exclusion by employing quantitative analysis of longitudinal panel surveys and/or qualitative methods to follow the dynamics of change. For example, were those individuals found to be experiencing poverty in rural England in 1980 the same people identified in a survey in 1990? Were we dealing with short spells of poverty experienced by many in rural society, or long spells experienced only by a small minority? This is of fundamental importance not only in terms of individual strategies but also in terms of the degree of solidarity within society.

As part of the JRF programme, one study followed the same randomly-selected 7,164 individuals each year between 1991-96, to help answer these questions (Chapman *et al.*, 1998). The results suggest that proportionately fewer individuals are affected by low income in rural areas, and that spells of low income tend to be shorter with a significantly lower proportion of 'persistently poor'. Despite this favourable comparison, prosperity is far from universal in rural Britain: a third of individuals in rural areas experienced at least one spell when their income fell below half the mean income, and 54 per cent experienced a spell with income below three-quarters of the mean income during these five years. The over 60 age group was significantly more likely to suffer persistent low income. Moreover, gross income inequalities intensified in both rural and non-rural areas over that period, which was characterized by major economic restructuring and cuts in public spending.

The analysis also confirms that the relative prosperity of rural households in Britain is not so much the result of strong rural economies but rather of selective migration. Richer people are moving into, and poorer people are moving out of, rural areas so causing a progressive gentrification of the countryside. Far from showing that rural people are part of an increasingly prosperous 'one nation', rising rural prosperity is an indication of an increasing spatial divide within Britain. Related research by Bate, *et al.*, (2000) confirms that there is a socially-selective and age-selective drift out of the towns and cities to the suburbs and rural areas, with only the relatively wealthy achieving the widespread dream of a house in the country. The less well-off can only move to the outer or inner suburbs, or must remain in the inner city. This issue, and the power relations which underlie it, are discussed below.

Low overall incomes in rural areas have often been blamed on limited individual pay, related to small-firms, lack of unionization, and a low skillbase. The research found only a weak relationship between low overall income and low individual pay, and far greater association between poverty and detachment from labour markets, despite the low levels of registered unemployment. Few of those on low incomes in rural areas are low paid, because few are in work. The greatest number is older people (see below). Of those of working age on low income in rural Britain, only 22 per cent are in employment; 23 per cent are self-employed (far more than in non-rural areas); 13 per cent are unemployed; and 41 per cent are detached from the labour market in other ways (such as the long-term sick [male] or family carers [female]). The composition of low-income households differed significantly between rural and non-rural areas with, for example, the self-employed a much more significant component of rural low income households than is found in non-rural areas. The processes behind these statistics are discussed in the next section, but it can be seen that they derive from the global penetration of local labour and product markets, state privatization and deregulation.

One of the most striking findings of the BHPS analysis (Chapman *et al.,* 1998) is how many of those on low incomes in rural areas are beyond working age and reliant solely on the state pension. The level of the pension is therefore of overwhelming importance to their income levels and to their quality of life. In addition, a special effort is required to reach people relying only on state pensions and unaware or unconvinced of their welfare entitlements, and to inform them of these in a sensitive and appropriate way. Specific policy changes also impact adversely on elderly people in rural areas, for example, via increased fuel prices and the diversion of business from small post offices. This social group is highly reliant on state systems, and (to a decreasing extent) on friends and family.

The most challenging finding of the research on disadvantage in rural Scotland was that rural people's own assessment was often at odds with official definitions of poverty (Shucksmith *et al.,* 1994; 1996). Most reviewed the improvements since their own childhood, when they lacked running water, electricity and televisions, and so could not conceive of themselves as poor. This is reinforced by constructions of the rural idyll. This has implications in considering ways in which such disadvantage can be corrected, both in terms of attempts at empowerment, and how to encourage people to take-up their benefit entitlements without stigma or loss of self-esteem. Overcoming resistance to these entitlements is a fundamental task for those seeking to tackle social exclusion.

Employment and Labour Market Integration

According to Berghman (1995), the three major 'bridges' towards inclusion are gaining employment, changes in family or household composition, and receiving welfare benefits, but are these the same in rural areas? And what particular constraints or 'barriers' are imposed by a rural context? Most poor people seek a full-time job as a route out of poverty, although this mode of escape is denied to many on the basis of their age, lack of skills, or childcare commitments. Are there additional obstacles facing those in rural areas, related to their small community, or the distances involved, perhaps?

Chapman *et al.,* (1998) found significant rural/non-rural differences in the demographic and economic events associated with escape from and entry into low income. Far fewer exits from rural poverty are associated with Berghman's three 'bridges'. This distinctive pattern applies also to entry into poverty, which is associated far less in rural areas with a fall in the number of earners (eg. following job loss or pregnancy), a change in family economic status (eg. retirement), or a change in household composition (eg. marital breakdown). This raises the question of what other 'triggers' and 'trampolines' may operate in rural areas.

Low pay is a particular problem. Persistent unemployment is less common but persistent low pay is more widespread in rural than in non-rural areas (Chapman *et al.,* 1998). The relatively limited 'escape rate' from low pay for individuals employed in small workplaces, combined with their dominance in rural employment, suggests that a lack of mobility from microbusinesses in rural areas may be an important explanatory factor. This was confirmed in the qualitative work by Monk and colleagues (1999) who found the following additional barriers to labour market participation:

- Structure of local labour markets – mismatches between jobs and skills;
- Employers' behaviour and attitudes – recruitment through informal social networks;
- Inaccessibility between home and workplace, and especially car-dependency; and
- Costs of participating in the labour market – childcare, eldercare and the benefits trap.

For some, integration into paid employment can resolve their poverty, perhaps with help from the extension of the New Deal to people over 50 together with related policy initiatives directed at transport, childcare and eldercare services. For others, it is the level and take-up of state benefits which offers the only prospect of escaping low income.

Work by Beatty and Fothergill (1997;1999) for the Rural Development Commission has found evidence of substantial hidden unemployment in rural areas, especially among men. Much of this took the form of premature early retirement and, in particular, a diversion from unemployment to long-term sickness. Distinctively rural dimensions to the problem of joblessness included the difficulties of 'getting to work', the narrow range of jobs available, the low level of wages on offer, and ageism among employers. More recently these authors (Breeze *et al.*, 2000) have investigated the operation of the New Deal programme in rural circumstances. Their principal conclusion is that while the New Deal addresses the supply side of labour market integration, it is demand-side problems (such as a lack of jobs) that remain deeply entrenched in rural labour markets.

Housing

The supply of affordable rural housing, whether through market, state, voluntary or kinship systems, has long been identified as essential to the vitality and sustainability of rural communities. It is also crucial to the life chances of many of the less prosperous members of rural societies, and therefore social inclusion. Yet, affordable housing is sadly lacking in many rural areas of Britain. The Countryside Agency and many others have identified the lack of affordable housing as *the* most important issue facing rural communities in England, and there is also compelling evidence from Scotland and Wales.

This is confirmed by Chapman *et al.*, (1998) who, as noted above, observed progressive gentrification of rural areas as the more affluent dominate the housing market. To understand better the lack of affordable housing in rural Britain, and the related social exclusion and social changes, one needs to consider the nature of, and influences on, the demand, supply and stock of housing in rural areas, and the roles of all four systems of market, state, voluntary social organisations, and family and friends. While there are important variations from one area to another in the ways in which these forces operate (documented by Shucksmith *et al.*, 1995) it is possible to summarise the general position.

Throughout Britain, the growing number of single person households and the increase in elderly people living apart from their families has increased the demand for housing. The demand in rural areas is augmented by the desire of many town-dwellers for a house in the country. At the same time, supply restrictions (notably planning controls) have permitted relatively few to realise the widespread desire for rural home ownership, and the resulting increase of house prices has caused problems for a sizeable proportion of the local rural population and for potential low income rural dwellers. House prices are now higher in rural areas than urban (Countryside Agency, 2001). Few new households in rural areas are able to afford home ownership through the open market.

As Newby (1985) elaborated, several years before the term social exclusion was coined:

> As prices inexorably rise, so the population which actually achieves its goal of a house in the country becomes more socially selective. Planning controls on rural housing have become – in effect if not in intent – instruments of social exclusivity.

The planning process has thus become the arena for a political conflict between those who favour countryside protection and those who seek 'village homes for village people' – something that has become more acute in recent years. Paradoxically, it may be that those most avidly protecting (their own) perception of the 'rural idyll' are, as a result of the effect on the housing market, threatening sustainable livelihoods in rural areas. In this way the operation of state systems of bureaucracy and authority, manipulated by powerful interest groups, works through housing markets to systematically force up house prices and thus exclude less wealthy households from many rural areas. These trends are likely to continue (PIEDA 1998).

Very little private housing in Britain is rented, and research suggests that this stock is unlikely to increase. As a result, the vast majority of those unable to afford house purchase in rural Britain must depend on social housing provision by the voluntary sector (housing associations) and local authorities. In each case this is allocated according to assessed need. However, social housing in rural areas is lacking, accommodating only 15 per cent of households. Partly this is a historical legacy of the dominance of rural areas by conservative councils who tended not to build council houses to the same extent; partly it is the result of social housing investment being concentrated in urban areas by the state bodies which finance voluntary housing; and partly it is a result of the Conservative policy during the 1980s and 1990s of selling council houses to tenants at substantial discounts, so privatising the former social housing stock. This clearly privileges those with ability to pay to the exclusion of those who exhibit housing need.

A number of studies have found that the problems of affordability in rural areas have worsened over the last decade. One clear reason for this has been the substantial shortfall of provision of social housing. Compared with an estimate that 80,000 affordable homes were needed in rural England between 1990-95, from 1990-97 only 17,700 new social housing units were provided (RDC, 1999). Even this contribution was offset by continuing discounted sales of social housing to sitting tenants.

Pavis *et al.,* (2000), echoing the other studies in the JRF programme, found that the young people they studied 'were neither wealthy enough to buy, nor were they poor enough to qualify for, the limited public sector provision'. One result of these difficulties is delayed household formation, with by far the majority of young people in rural areas, in contrast to elsewhere in the UK, remaining in the parental home. Although most were initially happy living with their parents, close to friends and family, problems became apparent later as they sought to assert their independence or when they found partners. At this stage their local housing opportunities were so limited that they had to leave, and Rugg and Jones (1999) found that 'almost all ended up living in urban areas'.

Studies have shown the very limited opportunities for affordable housing in most rural areas and that the interaction of state and voluntary systems may rupture kinship and friendship networks. For a fortunate few, social housing enabled them to stay within a particular village where they had lived for some time or had kinship ties. Respondents emphasized the importance of social networks in providing an opportunity to go to work while friends or relatives took on childcare responsibilities. This illustrates also how state and voluntary systems can work together with friends and family networks to redress the effects of market processes, so ameliorating exclusion.

Conclusion: Social Exclusion and Sustainable Livelihoods

In relation to market processes, this chapter has highlighted the barriers faced by those seeking integration into changing rural labour markets, and especially the shortage of well-paid, better quality jobs. In the course of globalization, international capital seeks to exploit those rural areas characterized by low wages, a non-unionized workforce, and lower levels of regulation, leading to increased casualization and job insecurity. This necessarily causes exclusion for some (eg, on the basis of age, lack of social connections or credentials). Other rural areas, and other individuals, are able to compete on the basis of quality through continuous innovation, cultural and social capital, and enjoy greater power and command over resources. This is one instance of the intersection of history and biography which this chapter has set out to explore, as market forces hold greater sway in relation to individual lives and life-chances.

Another illustration of this may be found in the difficulties many face in finding affordable housing, whether through market or state, to such an extent that they may be spatially excluded from living in many rural areas of Britain. The voluntary sector has been placed under increasing pressure as a result, while also becoming steadily incorporated into state systems though reliance on state funding and new forms of regulation. These intersecting spheres of social exclusion in turn have consequences for kinship networks and social support, as young people have to move away in search of affordable housing, higher education and better-paid employment. In these ways different dimensions of social exclusion interact to reinforce inequalities within rural areas, and between rural and urban areas.

It is important that, in analysing these processes one by one, we do not neglect the joined-up experience of each person's life. Accounts of people's lives gathered in rural Scotland illustrate how markets, state, voluntary systems and family and friends intertwine in complex ways within individual lived experiences, generating inclusion or exclusion (Shucksmith *et al.*, 1996). Car dependency, market and state allocation of housing, labour markets, education, training, childcare, family and friends are all relevant to people living in rural Scotland, and structure their choices and their livelihoods.

Perhaps most interestingly, the effects on individuals can be seen of the ascendancy of market processes, and the waning of state systems, as a result of the neoliberal hegemony which has hastened deregulation, privatization, reductions in public expenditure and global capital's penetration of labour and product markets. These effects vary from place to place, and from person to person, but in rural Britain a substantial number face social exclusion as a result – whether from casualization and job insecurity, from eroded pensions, from blurred accountability of agents of governance, or from delayed household formation and a lack of access to affordable housing. These changes in market and state systems also place considerable strain on voluntary systems, for example through feminization of the workforce and through additional reliance on volunteers, and on friendship and kinship networks. One high priority for future research should be to elaborate how the changing relative importance of these systems impinges differentially on people's realization of sustainable livelihoods.

Notes

1   See the April 2000 special issue of *Sociologia Ruralis*, Vol. 40(2).
2   See Philip and Shucksmith (1999).
3   Shortall (1999, p. 32) has pointed out that this often permits exploitation of women in farm families, since such reciprocity exists alongside a very unequal relationship of

economic and social power. Indeed, she argues 'one of the shortcomings of Polanyi's concept of reciprocity is its lack of any perspective on power'.

References

Bate, R., Best, R. and Holmans, A. (eds)(2000), *On the Move: The Housing Consequences of Migration*, York Publishing Services, York.

Beatty, C. and Fothergill, S. (1997), *Unemployment and the Labour Market in RDAs*, Rural Research Report 30, Rural Development Commission, Salisbury.

Beatty, C. and Fothergill, S. (1999), *Labour Market Detachment in Rural England*, Rural Research Report 40, Rural Development Commission, Salisbury.

Berghman, J. (1995), *Social Exclusion In Europe: Policy Context and Analytical Framework*, in G. Room (ed.), *Beyond the Threshold: The Measurement and Analysis of Social Exclusion*, Policy Press, Bristol, pp. 10-28.

Bramley, G., Lancaster, S. and Gordon, D. (2000), 'Benefit Take-up and the Geography of Poverty in Scotland', *Regional Studies*, vol. 34(6), pp. 507-520.

Breeze, J., Fothergill, S. and Macmillan, R. (2000), *The New Deal in a Rural Context*, Report to the Countryside Agency, Wetherby.

Byrne, D. (1999), *Social Exclusion*, Open University Press, Milton Keynes.

Cabinet Office. (2000), *Sharing in the Nation's Prosperity*, Stationery Office, London.

Cartmel, F. and Furlong, A. (2000), *Youth Unemployment in Rural Areas*, York Publishing Services, York.

Chapman, P., Phimister, E., Shucksmith, M., Upward, R. and Vera-Toscano, E. (1998), *Poverty and Exclusion in Rural Britain: The Dynamics of Low Income and Employment*, York Publishing Services, York.

Cloke, P., Milbourne, P. and Thomas, C. (1994), *Lifestyles in Rural England*, Rural Development Commission Rural Research Report 18, Salisbury.

Countryside Agency (2001), *State of the Countryside 2001*, Countryside Agency, Wetherby.

Duffy, K. (1995), *Social Exclusion and Human Dignity in Europe*, The Council of Europe, Strasbourg.

European Commission (1997), *Rural Development*, European Commission, Brussels.

Fiske, A. (1991), *Structures of Social Life*, Free Press, NY.

Giddens, A. (1991), *Modernity and Self-identity: Self and Society in the Late Modern Age*, Polity Press, Cambridge.

Gordon, D., Adelman, L., Ashworth, K., Bradshaw, J., Levitas, R., Middleton, S., Pantazis, C., Patsios, D., Payne, S., Townsend, P. and Williams, J. (2000), *Poverty and Social Exclusion in Britain*, Joseph Rowntree Foundation, York.

Hall, P., Gracey, H., Drewitt, R. and Thomas, R. (1974), *The Containment of Urban England*, Allen and Unwin, London.

Kesteloot, K. (1998), 'The Geography of Deprivation in Brussels and Local Development Strategies', in S. Musterd and W. Ostendorf (eds) *Urban Segregation and the Welfare State: Inequality and Exclusion in Western Cities*, Routledge, London, pp. 126-147.

Levitas, R. (1998), *The Inclusive Society? Social Exclusion and New Labour*, Macmillan, Basingstoke.

Meert, H. (1999), *Surviving on the Fringes of Society: Poor Rural Households in Belgium and a Typology of their Strategies*, paper to European Society of Rural Sociology XVIII Congress 1999, Lund, Sweden.

Monk, S., Dunn, J., Fitzgerald, M. and Hodge, I. (1999), *Finding Work in Rural Areas: Bridges and Barriers*, York Publishing Services, York.

Newby, H. (1985), *Green and Pleasant Land?*, Penguin, Harmondsworth.

Pavis, S., Platt, S. and Hubbard, G. (2000) *Social Exclusion and Insertion of Young People in Rural Areas*, York Publishing Services, York.

Philip, L. and Shucksmith, M. (1999), 'Conceptualising Social Exclusion', *Paper to European Society of Rural Sociology XVIII Congress*, Lund, Sweden.

Philips, M. (1993), 'Rural Gentrification and the Process of Class Colonization', *Journal of Rural Studies*, 9(2), pp. 123-140.

PIEDA, (1998), *The Nature of Demand for Rural Housing*, Department of the Environment, Transport and the Regions, London.

PIU (1999), *Rural Economies*, Cabinet Office, London.

Polanyi, K. (1944), *The Great Transformation*, Beacon Press, Boston.

Reimer, W. (1998), personal communication.

Room, G. (1995), 'Poverty and Social Exclusion: The New European Agenda for Policy and Research', in G. Room (ed.) *Beyond the Threshold: The Measurement and Analysis of Social Exclusion*, Policy Press, Bristol, pp. 1-9.

Rugg, J., and Jones, A. (1999), *Getting a Job, Finding a Home: Rural Youth Transitions*, Policy Press, York.

Rural Development Commission (RDC), (1999), *Rural Housing*, Rural Development Commission, Salisbury.

Scottish Office (1998), *Towards a Development Strategy for Rural Scotland*, The Stationery Office, Edinburgh.

Shortall, S. and Shucksmith, M. (1998), 'Integrated Rural Development in Practice: the Scottish Experience', *European Planning Studies*, vol. 6(1), pp. 73-88.

Shortall, S. (1999), *Property and Power: Women and Farming*, Macmillan, Basingstoke.

Shucksmith, M. (2000a), 'Endogenous Development, Social Capital and Social Inclusion', *Sociologia Ruralis*, vol. 40(2), pp. 208-218.

Shucksmith, M. (2000b), *Exclusive Countryside? Social Inclusion and Regeneration in Rural Britain*, Joseph Rowntree Foundation.

Shucksmith, M. and Chapman, P. (1998), 'Rural Development and Social Exclusion', *Sociologia Ruralis*, vol. 38(2), pp. 225-242.

Shucksmith, M., Chapman, P., Clark, G. with Black, S. and Conway, E. (1994), *Disadvantage in Rural Scotland: A Summary Report*, Rural Forum, Perth.

Shucksmith, M., Henderson, M., Raybold, S., Coombes, M. and Wong, C. (1995), *A Classification of Rural Housing Markets in England*, Department of the Environment Housing Research Report, HMSO, London.

Shucksmith, M., Chapman, P., Clark, G. with Black, S. and Conway, E. (1996), *Rural Scotland Today: The Best of Both Worlds?*, Avebury, Aldershot.

Chapter 14

Countering Localized Impacts of Globalization: Some Rural Community Development Initiatives in Australia

Erica Hallebone, Mary Mahoney and Mardie Townsend

Introduction

Definitionally, 'globalization' and 'localization' are individually contested, as is the relationship between the concepts: however, both concepts are helpful and relevant to the interpretation of recent and externally-generated socio-economic changes impacting on local regions. Voizey and O'Riordan (2001, p.26) note that:

> globalization through popular interpretations (journalistic and media representations) is a process of primarily economic, but also social and political change that encompasses the planet, resulting in greater homogeneity, hybridization and interdependence a global enmeshment of money, people, images, values and ideas that has entailed smoother and swifter flows across national boundaries. These processes are driven by technological advance, the growth of the informational sector, international cooperation, and processes of structural adjustment to a new global capitalist economic and political order headed by multinational corporations and international governmental institutions.

They also note the association with this concept of globalization of ideas of the promotion and domination of Western culture and capitalism to the exclusion of other cultures and economic systems, a loss of social diversity and the disappearance of local distinctiveness and community in favour of global culture and society (Voizey and O'Riordan, 2001, p.26).

As the concept of globalization ambiguously covers a broad canvas and is related also to recent debates about societal epochal change (eg. 'modernity', 'second modernity' and 'post-modernity'), it is of limited utility in assisting the interpretation of individual experiences in society, without placement in a more generalising frame of reference. As Voizey and O'Riordan point out, theoretically there are many precursors to the concept of globalization (for example, three types of theoretical framing are: in world-systems theory; globalization as an outcome of modernity; and globalization as a dual process centered upon culture) (Voizey and O'Riordan, 2001, p.28). Globalization as an outcome of 'modernity' is compatible with our own research traditions and present focus.

Giddens (1991), in an outcome of modernity focus, believed globalization could be best understood as expressing fundamental aspects of time-space distanciation – that is, conditions under which time and space are organized. Globalization concerns the intersection of presence and absence, and the interlacing of social events and social relations 'at distance' with local contextualities. Voizey and O'Riordan consider that globalization has a number of characteristics that are neither negative nor positive. It is not about the destruction of the locality in favour of greater homogeneity, but is about the greater interconnection and relativism between localities, economies, politics and cultures. Localization can be seen as an adaptation of these processes, a reaffirmation of what is local in the face of other localities. Importantly, localization is not a struggle to counter globalization; rather, it is the restatement of identity within a structure of multiple identities. They also believe that it is only at a local level that globalization has policy relevance. It is only here that a particular effect (such as the impact on employment) is perceived (Voizey and O'Riordan, 2001, p.41).

Other relevant societal theoretical concepts include what Beck (1999) referred to as 'second modernity', in which social relations based on the territorial sense of communities linked with the nation-state have changed. He describes the collective pattern of life progress and controllability, full employment and the exploitation of nature, as being typical of 'first modernity'. This is being undermined by five interlinked global processes: globalization, individualization, gender revolution, underemployment and global risks (Beck, 1999, p.2). Beck's analysis includes conceptualization of risk as a positive phenomenon. The communal sharing of risk, while not necessarily on a territorial basis, can be a (post-national) framework of reconstruction. He emphasizes the formula: 'labour is local, capital is global' (Beck, 2000, p.28). His perspective fits with globalization in the paradigm of second modernity in which 'globalization' is no longer understood as external and additive, but replaces the 'container image' of society and the state. It designates a transnational, despatialized power game, whose rules and boundaries, paradoxes and dilemmas, first have to be deciphered. Already we can see the outlines of the paradox of social closeness and geographical distance: that is, geographically remote differences and oppositions are lived and experienced as socially proximate, whereas geographical proximity leaves untouched the 'differentness' of social worlds. 'People can exist in one and the same place as if they were on different planets, while continents merge into a single social space in which people may live together across vast distances. The paradox of social proximity and geographical distance thus takes shape in a social-spatial figure: local disintegration and global integration' (Beck, 2000, pp.28-29).

Confirming that localization is presupposed by globalization, Beck anticipates the need for 'a service society tied to the locality' (Beck, 2000, pp.45). Export-intensive sectors of the economy are dependent not only upon the site of direct workplace interaction, but also upon a wider regional economy that includes insurance representatives, financial advisors, employees in snack bars or fitness centres, dentists and others who offer their wares for sale locally (Beck, 2000). The regime of risk entailed, sets off a 'dialectic of globalization and localization which are still not at all predictable in their consequences' (Beck, 2000, pp.73).

For individual 'destiny', it is relevant to reflect some of the ideas of Bauman (1998) who brought attention to local societal features, including mobility, which are both limiting and enabling as communal possibilities. As a stratifying factor in individuals' lives, the social construction of distance and freedom to move differentiates a global and extra-territorial elite from an ever more localized majority (Bauman, 1998, p.2). Individuals and families remaining in the localized communities bear the brunt of the problems and suffer anxieties and uncertainties, while those who move out (such as the young who move for education and those with marketable qualifications) experience the beneficial aspects of this change:

> What appears as globalization for some means localization for others; signalling a new freedom for some, upon many others it descends as an uninvited and cruel fate... Immobility is not a realistic option in a world of permanent change. And yet the effects of that new condition are radically unequal. Some of us become fully and truly 'global'; some are fixed in their 'locality' – a predicament neither pleasurable nor endurable in the world in which the 'globals' set the tone and compose the rules of the life game (Bauman, 1998, p.2)

He also makes observations that 'far from being hotbeds of communication, local populations are more like loose bunches of untied ends' (Bauman, 1998, p.24). Whereas the journeys of financial resources are as immaterial as the electronic network they travel, the local traces of their journeys are 'painfully tangible and real': 'qualitative depopulation', destruction of local economies once capable of sustaining their inhabitants and the exclusion of millions of people incapable of being absorbed by a new global economy (Bauman, 1998, p.75).

To summarize, there are two kinds of disadvantage of patterns of globalization to local communities: the global dictation of markets (in this instance, for agricultural products) and the loss of local infrastructure needed to sustain the community. Individual outcomes tend to depend on opportunities for personal mobility.

The second area of knowledge relevant to this study, historical-political action, concerns the implications of recent socio-economic and demographic changes in rural areas of Australia perceived (or not perceived) by state and federal governments and political parties in general. While recent Federal Government statements have assured rural Australians that service provision in rural communities will be maintained at present levels, for many communities the damage has already been done (Gray and Lawrence, 2001). If regional communities are to be viable in the future, many will need development assistance. As well, the success of community development initiatives in rural areas suffering socio-demographic and economic decline depends, in the first instance, on recognition of what has actually been lost, as well as on knowledge of the potential barriers or impediments to achieving such development.

Two case studies in a participatory social research project were conducted in rural communities in the State of Victoria, Australia, during 1998-99. The research was inspired by changes which have occurred in rural areas of Australia over the past two decades, in a context of policy changes at global, national and local levels, and political neglect of the needs of electoral constituents in rural and remote areas.

A town in the far north west (the Mallee region) and a second town in the south west of Victoria (the Western District) were included in the case studies which examined the cascading effects of economic and social policy changes on residents in the towns (Hallebone *et al.*, 2000).

Rural Australian residents, when compared with those of coastal and metropolitan cities, are generally more disadvantaged (Gray and Lawrence, 2001). Rural towns in Victoria are faced with changing demographic, social and economic profiles. Communities are experiencing population decline with young people moving away leaving an ageing and smaller population with diminished political power. The problems posed by these changes are compounded by the notion of available options to rural communities being limited predominantly to agricultural productivity. The combination of these factors leads to a loss of stability within rural towns.

The wellbeing of entire communities is under threat from the cumulative effects of policies that have sought to rationalize and centralize services. Attempts made by community members to sustain their communities are largely frustrated by external policy and other structural arrangements.

Socio-demographic and Policy Context of the Two Communities

Over the past two decades, in four out of every 10 rural shires in Australia, the population has declined, despite an increasing national population. In Yarriambiack, the shire in the Victorian Mallee 'wheatbelt' in which the first town is located, population loss has been 23 per cent. (The adjoining shire has recorded a loss of 34 per cent.) Several larger regional centres have absorbed some of the population losses of smaller towns. New farming technology, with resulting aggregation of farms and worker redundancy, has been cited as the major reason for population decline (Mitchell, 1999). Associated concern about the loss of services in rural and remote areas in Australia has led to consideration as to whether international human rights obligations are being breached by governmental failure to provide basic services such as health and education (MacDonald, 1999).

Researchers have noted that Australian rural residents (in a highly urbanized nation) suffer significant disadvantage (Cheers, 1990; Lawrence and Share, 1993; Walmsley, 1993; McLean, 1998). Rural dwellers overall have a lower wellbeing than their urban counterparts. This is associated with lower average incomes, higher prices, limited job opportunities, scarcity of health, childcare and other support services and limited educational opportunities (James, 1989; Cheers, 1990; Healthsharing Women, 1991).

In addition to the effects of globalization of agricultural markets and technical developments in agriculture, especially on rural youth employment, there is evidence to suggest that recent changes in public and private sector policies are compounding this rural disadvantage (Pritchard and McManus, 2000; Gray and Lawrence, 2001). Significant among the policy changes have been deregulation of the banking industry (resulting in the rationalization of banking services in rural areas), amalgamations in local government which further reduce local employment opportunities, compulsory competitive tendering in local government, the introduction of casemix funding for health services (resulting in closure of some hospitals and reduction of the range of health services in other communities), regionalization of government and privatized former government services, and closure of schools. An example of the implications of such policy changes is highlighted in a report prepared by the Productivity Commission (1999), which cites the example of rural railway workers in Western Australia who were made redundant because of contracting out of maintenance work. Their chances of finding work locally within a year of being made redundant were found to be less than 44 per cent, resulting in exacerbation of the problem of out-migration from rural communities (Pritchard and McManus, 2000). Cumulatively, the problems that these policies and practices pose for the social sustainability of the community are widely recognized (e.g. Cribb, 1987; Healthsharing Women, 1991; Jones and Tonts, 1995; Smailes, 1995; Steinberg, 1995; Safe, 1996; Winkler, 1998).

Cheers (1994, p.131) saw rural people as generally disadvantaged relative to urban Australians with respect to life chances, material and social deprivation, poverty, income levels, prices, unemployment, housing quality, education expectations and attainment, and occupation opportunities and achievement. He thought their total wellbeing had been taking 'a severe battering'. Cheers also noted (1994, p.131) that 'one third of all teenage suicides in Australia occur in rural areas (and) rural youth suicide rates are far higher than urban rates.' Despite this dramatic evidence, to date no previous research has provided an analysis of the cumulative impacts of these changes on individual, family and community health and well being in rural areas.

For most of this century, as Tonts and Jones (1997) point out, Australian governments' rural policies have been directed at providing a viable social and economic base for rural communities and country towns. Policies sought to ensure the conditions for stable growth in agricultural economies, with equitable levels of access to services such as schools, hospitals and public housing seen as instrumental in the development of stable rural communities. The recent process of global economic restructuring, agricultural readjustment, farm amalgamation and rapid technological change have contributed to the declining socio-economic viability of many rural areas including the case study communities. Continuing low commodity export prices are significant and the rising costs of domestic inputs into agriculture causes declining viability of rural communities.

Rural restructuring means fundamental change to the economic basis of society, from which radical social change ensues (Gray, 1994, p.17). While discussion about rural restructuring is often confused by apparent contradictions between long and short term trends and different interpretations of economic statistics, there is little argument that the long term economic trend for Australian agriculture has been one of declining returns and a decline in agriculture's share of the production of the nation's wealth.

Structural adjustment for people in many rural communities means a life which is frequently characterized by: poorly performing local economies, (being the product of drought, changing local and global markets, and low commodity prices); declining employment opportunities; low and often inadequate household incomes; out-migration of youth; negative health outcomes associated with stress and low incomes; increasingly mobile populations; increasing isolation in instances where farm sizes increase to maintain viability; and reduced access to services resulting from public service rationalizations (Fitzgerald, 1996, p.42).

This pattern of decline has been compounded by state and federal government policies based on principles of neoliberalism and privatization with increasing government withdrawal from regional development strategies (Gray and Lawrence, 2001). The levels of public service provision have been rationalized and responsibility for community wellbeing has been devolved to the local level, at the same time as the local level is being stripped of its capacity to cope, lacking the necessary competence or skills to deal with the changes.

Two Case Studies

The aims of this empirical project were to identify policy changes affecting the selected communities; assess the impacts (positive and negative) of those policy changes on the availability of services in the communities; assess the impacts of policy change and associated levels of service provision on perceptions of individual, family and community health and well-being; and identify the factors underlying any differences between communities in their perceptions of the impacts of the policy changes as well as record what developments were being undertaken. A research process based on inclusion and widespread consultation was adopted, to enable all interested community members to express their views. There was a high level of interest in and commitment to the study by members of the communities. Selection of the two communities in this study was based on Smailes' view that 'strongly defined communities of identity are centred upon small country towns with 'urban' populations of roughly 500 to 1500 people' (Smailes, 1995, p.145).

The study commenced in late 1998 and continued in 1999, and was undertaken in the two small rural communities which are of a similar size and geographic distance from the main Victorian metropolitan area (Melbourne). The findings of both communities showed very similar trends.

The Mallee town is located four and a half hours by road transport from Melbourne and two and a half hours from the nearest regional city. It has a population of 670 people, with 93 per cent being Australian born. Of this population 81 per cent are aged 15 and over and 29 per cent of the population are over 65 years of age. The unemployment rate is some 4 per cent and 40 per cent of the population is not in the workforce (Australian Bureau of Statistics, 1996).

The Western District town is located approximately three hours by road transport from Melbourne and half an hour from the nearest regional city and has a population of 503 people, with 86 per cent being Australian born. Of this population 78 per cent are aged 15 and over and more than a quarter of the population are over 65 years of age, with more than 15 per cent aged 75 and over. There is an unemployment rate of some 4 per cent with 42 per cent of the population not in the workforce (Australian Bureau of Statistics, 1996).

In both communities, key informant interviews were conducted with relevant community representatives and stakeholders. Focus groups were also conducted with relevant groups such as young (teenage) residents, a parents' group, and meetings were held with the local community development associations (Progress Associations). Over 80 people contributed to the study.

Findings of the Case Studies

The following section briefly summarizes the residents' perceptions of the impacts in farming, local government, public utilities, banking, health care and education: all are interconnected (see Hallebone *et al.*, 2000).

Changes in Farming

In both communities, changes in government subsidies, transport costs, droughts, and sale of farms have caused relocation of farmers off the land and into the town. In the Western District town, changes in the floor-pricing of wool and a poor beef market have reduced the money coming into the community for reinvestment, with negative impacts on local businesses and employment.

Farms are becoming larger and people found it necessary to diversify their income bases. The decline in the viability of farms has also necessitated an increase in off-farm work. Results include the reduction in availability of volunteers for community services (including essential services such as fire safety) due to absence from the local community and/or the need for farm work to be done after hours; and the tendency to do business in the town near work. Together, these effects undermine the viability of businesses and voluntary organizations and make the communities less attractive places for other residents.

Merger of Shires

Centralization of local government services away from the communities caused the loss of thirty jobs in the Mallee town and seven in the Western District town with consequent losses of families, local businesses, voluntary contributions, and a higher load on remaining residents. Cascading impacts rebounded into support services with the downgrading of schools and banking facilities. Critical and significant losses expressed were of major focus (Shire office – originally one in each town), community identity and representation and a shift from feelings of self-satisfaction to feelings of apathy, despair and isolation.

Compulsory competitive tendering (CCT) in local government has compounded this, with local businesses not necessarily winning tenders. A further loss of twenty jobs was expected in the Western District town from the result of council staff being forced to tender for their work.

Service provision impacts from local government mergers and CCT (when still government policy) rebounded into the community. In the Western District town, the National Australia Bank branch closed soon after the amalgamation of shires, when over $1.5 million previously held by the branch was relocated to Hamilton, the administrative centre of the new Southern Grampians Shire. Residents perceived that the town has a low priority for funding within the new Shire.

Similarly, in the Mallee town, impacts at the local government level rebounded into support services. Feelings of despair included the ongoing threat to livelihoods of people still employed within the community by the Shire.

Downgrading of Public Utilities

Until recently the Mallee town provided infrastructural support within the region for services such as water, electricity, natural resources and environment. With the rationalization of these services additional costs were experienced for water supply (to fill the local pool and artificial lake), transport of freight and people, as well as frequent disruptions to supply. Loss of a base of the State Government Department of Natural Resources and Environment had a similar impact to the loss of local Shire offices. These changes have resulted in a loss of youth employment and training opportunities within the community.

Until the Shire amalgamation, the Western District town had its own library and its own water board. Removal of these services has contributed to local job losses. While some residents welcome the new mobile library service, others consider it an unsatisfactory replacement for the community's own library. Similar views apply to the downgrading of the ambulance service, and the privatized Post Office (though perceived as providing a good service) was seen as an example of lack of government commitment to providing services in rural areas.

Deregulation of Banks

Downgrading of banks has meant additional loss of (bank staff) families to the community. Residents must now travel outside the community to discuss critical financial affairs with regional branches leading to purchasing practices outside the community. There is a strong sense that people within the community are being dictated to by the banks on the nature of their banking, particularly through the nature of the services provided. Another impact is the loss of 'wisdom' within the communities. This loss of bank staff coupled with the loss of significant other professionals and clergy within both towns, means that the communities no longer have a pool of future elders, for example, probation officers or school councillors.

Health Care

Health delivery services in the two towns were chiefly characterized by losses. Rationalization of country health care provision has meant the merger of three hospitals within the region around the Mallee town to create one public hospital with emergency care facilities in the town. While very strong community involvement was required to secure the facility, in the face of various losses experienced by the residents, it provides a sense of some future for the town. To residents, it also represents 'the only fight we've won.'

Relocation of Home and Community Care (HACC) services out of both communities to the nearest regional towns half to one hour away, means that support to elderly residents has moved to outside locations resulting in depersonalization and loss of local knowledge. Relocation of the HACC worker has caused loss of income to the towns and a trend towards reliance on trades and services outside the local communities. Dental and pharmaceutical services have largely been lost and controversies arisen about absolute availability of any medical practitioners in the respective towns.

Changes over recent years in the funding arrangements for hospitals and health services in the Western District town have affected health care services. With the change to casemix funding, its hospital experienced an increase in funding due to the categories of patient types it was serving. The move from casemix to 'Healthstreams' funding has provided greater flexibility in the use of funds, with money able to be shifted into the areas of greatest need. But while this creates a greater capacity to meet some of the community's health care needs (such as aged care and allied health services), this gain has been offset by a decline in acute care facilities.

Many concerns were expressed about the impacts of limited transport options, particularly on the ageing members of the populations of both towns. Volunteer drivers provide transport for those who cannot get themselves to the larger towns for appointments, but for an ageing population this will become more and more problematic. As well, the amalgamation of local governments has seen the introduction of a 'user pays' policy for the local community bus services. Ambulance services have been reduced dramatically and the timing of bus services do not allow flexibility for residents to attend specialist appointments.

Education

Downgrading of the Mallee town secondary school due to declining population has meant reduction of overall staff numbers and has also led to less effective rotation of teaching staff. Loss of families has meant a shift in demographic profile of the community and fewer young people staying in the local community for their secondary education. Once young people leave the community they do not return because there are no employment prospects. This is not welcomed by young residents, many of whom appreciate the quality of life that the particular towns offer and want the chance to settle there after they leave school.

Country schools need to provide pastoral care for students. Increasing suicide rates amongst rural youth mean that there is extra pressure for resources. The Western District secondary college responded to this pressure by appointing a part-time chaplain to the school staff.

Changes to funding policy for pre-school education provision means that fund raising is now a requirement to provide for staff salaries, not an option. This is set in the context of low income and lack of permanency of jobs for families, yet the need to provide more and more for the education of one's children.

Residents' Perceptions

It was perceived that in terms of the capacity to engage in community development, residents' perceptions of the cumulative impacts of all these recent changes could be divided into four groups:

- The 'optimists', with a vision for the future, actively working towards the long term viability of their town: these appeared to be a minority;
- The 'pessimists' who believe that the town is doomed and that they are simply the next level to become a ghost town, or at best, has a future which is precarious;
- The 'apathetic' who seemed not to care or did not care; and
- The 'disenfranchized' who are not seen to belong to the respective community and are therefore operating outside discussions about the future.

While individual perceptions of the health and well-being of the communities obviously were influenced by the resident's individual stance or position, the sense of powerlessness against outside threats was very strongly articulated by every group.

Community Development Strategies

Progress Associations of both towns are planning and promoting various socio-economic and community development strategies. Community development is understood as being a process of involving people who live in a defined geographical locality, or who share common interests, in collective action to improve their well-being (Baum *et al.*, 1992). Among those residents in the two towns, those who tend to be 'optimists' see tourism as the best opportunity for various forms of development. Both towns are on the route to National Parks.

The Mallee town has built a 'picturesque' lake and recent successful grant applications have supported the construction of facilities within a 'creative village' concept. Without doubt, this success has given the residents the feeling that something positive is happening. Concern is expressed about appropriate signage and the importance of it controlling its own image rather than being dictated to by slow-moving external agencies. Access to the National Park is also crucial from a tourism perspective, especially the provision of an airstrip. It is also seen as advantageous from an emergency health care perspective.

The Western District town's location on the main route between the coast and the Grampians (National Park) was viewed as a positive. So were local features such as: an historical home; a German settlement of the area and the associated cultural features (music, cuisine); the volcanic crater; the wetlands development in the town; and historical events such as the town's Nineteenth Century bank robbery (which provided the focus for a recent re-enactment pageant). There was a general view that, if the town could capitalize on these features, its future might be brighter. However, the attitude of Southern Grampians Shire was seen as unhelpful in this regard, with poor signage of the Western District town and its attractions in the local area.

In the Western District town, the availability of the former Shire Offices and other buildings suitable for tourist purposes, wide roads, pleasant tree-lined streets, plentiful housing at reasonable prices, the natural spring (recently uncovered), and the general friendliness of the people are assets which the town can utilize as a basis for future development. In addition, with people who have recently lost their jobs as a result of the structural changes, and with new people moving into the town to take up (cheap) housing opportunities, there is, or potentially will be, a pool of people with a variety of skills which can be used to develop new enterprizes.

In the Mallee town the creation and success of an employment and training agency is seen as very important to the community. Its siting in the old Shire Offices is important in re-establishing that critical focus for the community. Local perceptions of what the agency is contributing to the community are very positive. An opportunity (perceived during the beginning of the research process) to appoint a Chief Executive Officer was seen as important, especially in the context of previous concerns about lack of a professional base within the community.

Another new development in the Mallee town was the planned location of a mobile telephone tower nearby being vital to the long-term viability of the town; for example, travelling business and other personnel in the town can now stay in the town overnight and still communicate with their organizations.

Perceived Barriers to Community Development Strategies

While the need and opportunities for community development strategies to be implemented in the case study communities is obvious, the current profiles of the communities indicate that there are a number of factors which impede such strategies. These reflect similar concerns to those found in the literature, and include: population size and profile; economic resource base; lack of cultural congruence; loss of leadership capacity; declining accessibility of information; and inappropriate government structures/requirements.

Population Size and Profile

Due to declining populations within the two case study communities, there are fewer individuals available to undertake community development strategies. As well, the age profile of both communities indicates that they have fewer young adult to middle aged residents (between 24 and 65) and more residents in the over 65 age group than either Victoria as a whole or Australia. While community development strategies will obviously need to involve those of all age groups, the higher than average elderly populations in these communities suggests that there may be a shortage of those with the energy to initiate and sustain such strategies.

Economic Resource Base

The economic resource base of these communities is declining, with a reduction over recent years in the number of businesses, the number of jobs and the average income. With government policies over recent years resulting in an increasing need for communities to undertake local fundraising in order to sustain community services and facilities, the capacity and the willingness of residents to dip further into their financial or 'in-kind' resources in order to undertake or sustain community development strategies has been compromized.

Lack of Internal Cultural Congruence

Local government mergers have caused some 'crossing of cultures', resulting to some extent in a lack of cultural congruence between these communities and their new local government bodies. In both communities, this has led to a perception by residents that their towns' development opportunities are being 'poached' by the large towns at the centre of their new shires. Within the communities, there is also a growing lack of cultural congruence, with the increasing number of vacant residential properties being sold or rented to people moving from the cities in search of cheap accommodation. These people are often not aware of the need for community involvement in sustaining rural towns, and this can result in misunderstandings between new residents and those of long standing.

Loss of Leadership Capacity

With farm aggregation, loss of banks and local government offices, loss of businesses, and the resultant declining employment opportunities, young people are moving away from these communities to study and are not returning. Similarly, those young families who previously would have moved to the area because of jobs in banks, schools, shire offices, government departments and the like are not doing so. This has resulted in a loss of leadership capacity in these communities, as these people (who tended to be those with educational qualifications or other training) are not available to share their skills.

Declining Accessibility of Information

Despite the perception of living in an 'information age', where information is more readily available than ever before, these communities are finding that information is only selectively available. Familiarity with technology is an essential component in being able to access information, and this is less readily available in these small communities than in larger centres. Moreover, the regionalization of government departments, and the privatization of many former government services, has further limited access to relevant information. One example given was the trend for information relevant to agriculture, once received from a large Department of Agriculture, being selectively distributed through private agribusinesses. For successful community development, information about potential strategies and funding sources are essential.

Inappropriate Government Structures or Requirements

Lack of continuity of government structures, policies and procedures contribute to the difficulties faced by small communities in undertaking community development, with community members feeling that they are being asked to reinvent the wheel each time they seek help. The human leadership capacity of these communities, already stretched because of the declining populations and diminishing economic base, is further undermined by having to deal with a plethora of government departments and regional structures, with no capacity for integration and information sharing. This is complicated further by differing timelines and requirements for grant applications, and accountability procedures that are often onerous. Small communities, which lack access to or understanding of information about the necessary procedures because of their existing disadvantage, are in a 'Catch 22' situation: they cannot access the grants/support because they do not have the knowledge or skills, because they have not been able to access grants to build up such knowledge or skills.

Conclusion

Community development initiatives in these two rural areas are highly frustrated not only by external neglect but also by disintegrated policies and sources of assistance. Experiences within these communities, (which may be similar to many others across Australia), show not only that strategies need to be put in place to assist them enhance community sense of well-being and survival but also – to be successful – community members need to be realistic.

There is little evidence of recognition in policy frameworks of issues of long-term sustainability for rural communities. To the extent that the trends of the agricultural industries, new technology, reduction of the welfare state and privatization tend to lead to a loss of community identity and increase in individualism, the chances to build local initiatives to deal with change are diminished.

However, Beck (1999), focussing on risk in society and its possible positive connotations, has drawn attention to a framework for future reconstruction of communities.

In proposing a trial of the Nebraska model of community revitalization to be applied to a Victorian town (Yarram) Worrall (1999, p.3) points to the need for new strategies for developing and maintaining a program of change management. Four basic elements are required: an appropriate model of change, a strategy, suitable leadership and a community committed to trailing the model.

In the context of globalizing capitalism and a national rural policy vacuum, the survival of communities such as these will depend almost entirely on the initiatives of new local industries and the residents' construction of networking and social capital. For both types of initiative, in an era of the accelerating individualism nurtured by globalization, enhanced personal and group energy will be required.

The very existence of small rural communities is crucial to the sustainability of Australia, economically, socially and bio-physically. It is to this end that methodologies which promote interaction and linkages between policies, legislation, education, partnership and advocacy arrangements must be encouraged (Labonte, 1993). Onyx (1997, p.2) recommended future research be conducted within an action framework to focus on community possibilities of developing social capital: forms of mutual participation in collective self management.

As Gray *et al.* (1995) noted, the issues of empowerment and equity need to be considered within society relative to integrating economic, social and physical imperatives, and to cater both for those individuals who have access to resources and those who do not. At present, not much is known about farmers' perceptions, for example, whether they view reflexively their situation within global capitalism or see the new productivity demands, like weather and the market, as being part and parcel of agricultural life (Gray *et al.*, 1995). This converges with Beck's contemporary gathering of views of communities at risk being potential bases of political action and collectively-binding decision-making in a 'second modernity' (Beck, 1999). For this to become a greater possibility, recognition should be given to what has been lost (and gained) in rural communities. Research information on the impacts of change should be made available to community groups, together with globally-based knowledge of possible development strategies. This must include potential barriers to community development and how those barriers can be broken.

References

Australian Bureau of Statistics. (1996), *Australian Demographic Statistics 1996 (Census Edition, December Quarter)*, Catalogue No. 3101.0, Australian Commonwealth Government Printer, Canberra.

Barkley, D., Smith, S. and Coupal, R. (1991), 'High Tech Entrepreneurs in Small Towns', *Journal of the Community Development Society*, vol. 22(1), pp. 38-55.

Baum, F., Fry, D. and Lennie, I. (eds)(1992), *Community Health: Policy and Practice in Australia*, Pluto Press, Sydney.

Bauman, Z. (1998), *Globalization: The Human Consequences*, Polity Press, Oxford.

Beck, U. (1999), *World Risk Society*, Polity Press, Cambridge.

Beck, U. (2000), *The Brave New World of Work,* Polity Press, Cambridge.

Cheers, B. (1990), 'Rural Disadvantage in Australia', *Australian Social Work*, March, vol. 43(1), pp. 5-13.

Cheers, B. (1994), 'Social Development as a Facet of Regional Development', in A. Bundy (ed.), *Proceedings of the First National Conference on Building a Better Future for Regional Australia*, South Australia, University of South Australia, Whyalla, pp. 54-60.

Cheers, B. (1995), *Integrating Social And Economic Development In Regional Australia*, Centre for Social and Welfare Research, James Cook University, North Queensland.

Cribb, J. (1987), 'Death of a Country Town', *The Australian*, 14-15 February, Supplement 1, p. 1.

Fitzgerald, R. (1996), 'Sustainable Development and Rural Structural Adjustment', in *Rural Communities Looking Ahead*, New South Wales Rural Social Policy Conference, Dubbo, June.

Giddens, A. (1991), *Modernity and Self-Identity: Self and Society in the Late Modern Age*, Polity Press, Cambridge.

Gray, I. (1994), 'The Changing Structure of Rural Communities', *Rural Society,* vol. 4(3/4), pp. 17-21.

Gray, I., and Lawrence, G. (2001), *A Future for Regional Australia: Escaping Global Misfortune*, Cambridge University Press, Cambridge.

Gray, I., Phillips, E., Ampt, P. and Dunn, T. (1995), 'Awareness or Beguilement? Farmers' Perceptions of Change', in P. Share, (ed.), *Communication and Culture in Rural Australia*, Centre for Rural Social Research, Key Papers Series, 4, pp. 53-69.

Hallebone, E., Townsend, M. and Mahoney, M. (2000), 'Globalising Policy Changes: Social Impacts in Two Australian Towns', *Journal of Family Studies,* vol. 6(2), pp. 214-230.

Healthsharing Women, (1991), 'Women, Health and the Rural Decline', *Healthsharing Women*, January/February, pp. 5-9.

James, K. (ed.)(1989), *Women in Rural Australia*, University of Queensland Press, St. Lucia.

Jones, R. and Tonts, M. (1995), 'Rural Reflections and Social Sustainability: Some Reflections on the Western Australian Wheatbelt', *Australian Geographer*, vol. 26(2), November, pp. 133-139.

Labonte, R. (1993), *Number 3 Health Promotion and Empowerment Frameworks*, Centre for Health Promotion, University of Toronto.

Lawrence, G., and Share, P. (1993), 'Rural Australia: Current Problems And Policy Directions', *Regional Journal of Social Issues*, vol. 27, pp. 3-9.

MacDonald, J. (1999), 'Basic Rights Denied in Bush: Report', *The Age*, Melbourne, April 28, p. 6.

McLean, R. (1998), Keynote Address, *Victorian Rural and Remote Allied Health Professionals' Conference*, Melbourne, May 29-31.

Mitchell, B., (1999), 'Farm Mergers Leading to Population Decline in Rural Areas', *The Age*, Melbourne, February 26, p. 3.

Onyx, J., (1997), How is Social Capital Being Measured? *Social Capital Conference, University of Queensland and Queensland University of Technology,* Brisbane.

O'Riordan, T. (ed.)(2001), *Globalism, Localism and Identity*, Earthscan Publications, London.

Pritchard, B. and McManus, P. (eds)(2000), *Land of Discontent: The Dynamics of Change in Rural and Regional Australia*, University of New South Wales Press, Sydney.

Productivity Commission. (2000), *Progress on Rail Reform* (Draft Report), AusIn, Canberra.

Safe, M. (1996), 'Trundling On – Trundle, Like Too Many Australian Country Towns, is Fighting for Life', *The Australian Magazine,* 2-3 November, pp. 10-15.

Smailes, P. (1995), 'The Enigma of Social Sustainability in Rural Australia', *Australian Geographer,* vol. 26(2) November, pp. 140-150.

Steinberg, M. (1995), 'Don't Just Assume We'll Cope: Implications for Community Care of Changes in Hospital/Community Interfaces, in Community Care... The Next 20 Years', *Proceedings of Carers of Australia Conference,* Canberra, 30-31 March, pp. 73-80.

Tonts, M. and Jones, R. (1997), 'From State Paternalism to Neoliberalism in Australian Rural Policy: Perspectives from the West Australian Wheatbelt', *Space and Polity,* vol. 1(2), pp. 171.

Voizey, H. and O'Riordan, T. (2001), 'Globalization and Localization', in T. O'Riordan, (ed.), *Globalism, Localism and Identity,* Earthscan Publications, London, pp. 25-42.

Walmsley, D. (1993), 'The Policy Environment', in T. Sorensen and R. Epps (eds), *Prospects for Policies for Rural Australia,* Longman Cheshire, Melbourne.

Winkler, T., (1998), 'Bank Closure Gives Town Withdrawal Symptoms', *The Age,* 6 May, p. 8.

Worrell, A., (1999), *Change Management in Rural Communities,* Victorian Farmers Federation, Melbourne.

Chapter 15

The Problem of Trust at the Margins of the Network Economy: Trust and Confidence among Rural Entrepreneurs in Finland

Petri Ruuskanen

Introduction

For several years now the need for increased flexibility in most aspects of industrial production has been one of the central themes in economic debate in the industrialized countries. 'The network enterprise' has become the ideal model of the competitive firm. This has resulted in the process of disintegration of large vertical firms and their transformation into global business networks (Castells, 1996). As a part of the debate, small firms have attracted the attention of governments' economic policy. It has been claimed that these developments give small enterprises a new competitive edge in the world economy. Piore and Sabel (1984) for example, have argued that the rise of new production models takes small enterprises and their co-operation to the centre of the flexible economy. At the same time flexible co-operation of small firms would make sustainable business activities possible in previously non-industrialized rural areas.

Therefore, it is not surprising that small-scale enterprises and industrial networking have also been brought forward as solutions to the problems of competitiveness and unemployment confronting Finnish rural areas. The primary goal of the current Finnish rural policy is to encourage co-operation between small rural enterprises, such as small sawmills, furniture makers and specialized food producers, and larger, medium-sized companies, which may already have experiences in export trade.

Building co-operative networks also requires the building of trust. Autonomous pursuit of self-interest is often less profitable to all parties involved, and if trust is absent in social networks, all actors will sacrifi.

the potential gains of co-operation to autonomous pursuit of self-interest. Furthermore it has been stated that trust makes networks of small firms successful and improves their business performance since it encourages the disclosure of information in networks and reduces uncertainty and the need for monitoring the activities of business partners (see Sako, 1992 and 1998; Powell, 1996).

In this chapter I shall focus on the ambivalent attitudes toward networking among rural entrepreneurs in Finland. The problem of trust and confidence will also be examined. The chapter is structured as follows: The following (second) section summarizes the different benefits of flexibility sought through networking. The third section provides a short overview of the problem of trust in flexible business networks. In the forth section I turn to the interpretation of my qualitative data. The results are based upon open-ended interviews conducted with self-employed entrepreneurs and small employers (with less than 10 employees) in rural Finland.[1] A brief discussion is found in the concluding section.

Networking and the Need for Flexibility

In today's world, the management of change and uncertainty is a central problem for most economic organizations. A rapidly changing market presupposes more flexibility and adaptability from economic organizations than ever, which has led large corporations to adopt new structures of management. Business networks are widely regarded as the best way to organize production in a volatile economy.

It has to be remembered, however, that there are many alternative ways of seeking flexibility through networking. In the ideal model of flexible specialization, as outlined by Piore and Sabel (1984), the strategy of constant innovation would enable small firms organized in loose horizontal networks and vertical chains of co-operation to compete with the mass production model dominated by large firms. On the one hand, small enterprises can work together to diversify their production and seek competitive advantage in narrow, specific market segments, or niches. On the other hand, it is possible to develop vertical networks (subcontracting by a core group of companies in the centres to smaller firms). Thus, collaboration can be based on either a horizontal or a vertical network structure.

For large manufacturing companies, for example, it may indeed be rational to outsource their peripheral tasks in order to seek out flexible reservoirs in the labour market. Atkinson (1987) distinguishes three goals for employers in seeking greater labour market flexibility:

- *Numerical flexibility*, which is concerned with how employers adjust the numbers of people employed or the hours they work to changing workloads;
- *Functional flexibility*, which is concerned with how they adjust the deployment of these people; and
- *Distancing*, which represents the replacement of employment contracts by commercial contracts such as subcontracting agreements.

Each type of flexibility impinges differently on different groups within the labour force, which as a result becomes segmented into a 'core' group of employees surrounded by 'peripheral' and 'external' groups.

A trade union can nevertheless restrain this flexibility within a firm, which makes outsourcing attractive, as small-scale entrepreneurs – self-employed persons and small employers – are supposedly more willing to accept flexible working hours, an intensification of work or a squeeze on their profits to deal with the fluctuations of the market. At the same time, outsourcing makes it possible for the larger firms to achieve more flexible financing by shifting part of the investment burden to subcontractors. It even seems that the conscious efforts of the core-periphery strategists favour the more systematic use of an external workforce rather than organizational restructuring within firms (Hakim, 1990).

Outsourcing and the segmentation of the labour market can give rise to new firms through a pull-push mechanism. On the one hand, the process opens up new business opportunities by creating a demand for small-scale suppliers, but on the other hand, it creates an 'unemployment push', which may force people outside the core labour market to establish small firms of their own. The latter seems to be very common among small rural entrepreneurs in Finland, especially where the smallest industrial firms are concerned (see Niittykangas, 1992, p.90). This double mechanism affecting the formation of new companies may also produce wider and sharper segmentation of the 'external' labour market than of the 'internal' labour market.

Flexible Networks – The Problem of Trust and Confidence

It is often stated that globalization process and more flexible use of labour leads to differentiation and individualization in the labour market (Beck, 1993; 1994). The individualization of the labour market is most striking at the margins of the network economy. Small-scale subcontractors, self-employed entrepreneurs and freelancers obviously have to 'produce, stage and cobble together their biographies' (Beck, 1994, p.13) and tackle risky opportunities offered by the flexible economy by themselves. Regarding individualization, it is quite fitting that the individual entrepreneur – and his or her spirit of enterprise – has become the symbol of the ideal actor in a flexible economy.

The themes of entrepreneurship and industrial networking have also been adopted as measures in the rural policy of the Finnish government – particularly as a response to unemployment. Today, it is often regarded in rural policy discourses that the prosperity of the rural population lies in people's own initiative and entrepreneurship. Most recent official documents on rural policy include statements in which rural inhabitants are encouraged to become self-employed entrepreneurs, subcontractors and distance workers. 'Resistance to change' and 'the prevailing public opinion' are, however, seen as obstacles to this development. The traditional cultural patterns shared by most rural entrepreneurs appear to be opposed to new, flexible production models such as networking (see CRP, 1996, p.109).

It has been suggested that even in modern industrialized societies economic behaviour is embedded in social relations. The argument of embeddedness signifies, basically, that economic organizations are not developed in a social vacuum, but economic activity is affected by the socially constructed institutions, the actors' personal relations and the structure of the network of relations (Granovetter, 1985; 1992). Social capital – social networks, norms, and trust – is today widely regarded as a prerequisite for successful collaboration and economic development (Coleman, 1988; Putnam *et al.*, 1993; Woolcock, 1998).

The performance of a given industrial network, for example, largely depends on its connectedness – that is, it's structural ability to facilitate communication between its components. Performance, however, also depends on consistency, such as the extent to which the network's goals and the goals of its components coincide (Castells, 1996, p.171). Analogously, social capital contains both a structural bridging dimension and a cohesive binding dimension.

The structure of the ties in the networks constitutes the bridging dimension of social capital. Trust, on the other hand, can be considered a central element in the social capital that binds social networks together. It is often stated in discussions concerning business networks that the further development of industrial networking requires the development of trust in the network (Sabel, 1989; Powell, 1996; Sako, 1998). The precise definition of trust, however, remains unclear and elusive. The term implies various degrees of hope, belief, presumption, expectation and so on. What is important with trust – in the sense it is discussed here – is that it must also have bearing on one's action (Dasgupta, 1988, p.51).

Trust can also be conceptualized as a public asset, which improves the capabilities of complex systems (such as the economy or the industrial network) by reducing contingencies and increasing predictability and stability. If one trusts the stability of the European monetary system, for example, then one is not likely to be preoccupied with protecting savings by buying US dollars. There is confidence enough to keep assets in a European bank account, rather than to invest abroad. Trusting action supports the stability of the currency and so strengthens trust further.

Trust is also (and obviously) an interpersonal social relationship. When a person trusts someone, there is some acceptance that at a future time that trust might be broken. If the person tries to 'hedge' or insure themselves against possible disappointment, then there is no trust. On a micro or a personal level then, trust is accepted dependency upon the actions of others (Lewis and Weigert, 1985, p.464), or accepted vulnerability to another's possible, but not expected, ill will (Baier, 1986, p.235).

To make the two aspects of trusting clearer, Niklas Luhmann (1979; 1988) and Adam Seligman (1997; 1998) suggest that we should distinguish between confidence and trust. Confidence is what a person has when he or she does not consider alternatives, but knows what to expect in a situation. In this sense confidence can be defined as an expectation which is based on knowledge that an interaction is set within a system or a context through which it is possible to impose sanctions in the case of a violation of an obligation or an agreement. The expectation can be based, for example, on previous experiences, knowledge, familiarity or some other thing that makes the outcome predictable. It brings a possible future into the present (Barbalet, 1998, p.87). Trust on a micro level, as noted above, refers to a situation with the reflected and accepted possibility to become disappointed. While the former is associated with known systems or contexts, the latter is usually associated with the margins of these familiar contexts and has to be constantly maintained by personal interaction.

According to Seligman (1997) interpersonal trust is an historical phenomenon which has been developing hand-in-hand with modern society and the risks connected with modern forms of social relations. In traditional societies the closure of social networks, effective norms and sanctions increased the predictability. They were societies with high levels of confidence. In modern urban societies a social structure is usually much more open. People often meet total strangers, whose actions are difficult to predict. Today people have to decide if they can trust others, their business contacts for example, as actors, who may be granted a certain degree of freedom to act on their own. Thus, trust may be seen as a modern complement to confidence, developed as a response to increasing individualization.

Trust and confidence are also to some extent dependent on each other. To be able to trust someone presupposes confidence, or in Giddens' (1991) terms 'basic trust'. If nothing can be taken for granted, it is very difficult to trust other individuals. In this sense, the consequence of the lack of confidence is not the development of trust or distrust, but feelings of uncertainty, anxiety, and dread. On the other hand, in a context of generalized confidence there is no need for personal trust, since everybody's action is predictable (and generalized trust is, on the other hand, confidence).

The requirement of trust is quite evident in the modern, volatile economy, where opportunities and market niches rapidly appear and disappear. The web of business networks and subcontracting agreements also has to change and adapt to market fluctuations. In a situation of rapidly changing relations, interdependent business partners may either be friends, strangers, or 'neither friends nor strangers' – as Lorenz (1988) descriptively entitled his article dealing with the issue. Therefore, the formation of network relations involving mutual dependency requires both of these dimensions of trusting action. An entrepreneur considering co-operation with other entrepreneurs has, first, to be confident that a networking system works. Equally, he or she has to be confident that all other networkers are competent in their own trades, and that his or her own role in the network is remunerative. At the same time a person must trust the goodwill of business partners, their trustworthiness as individuals, and that they will fulfil their obligations.

It is likely that if a person has proven to be untrustworthy in a private sphere, professional credibility will also be eroded. Let's take a plumber who (for any reason) is entrusted the care of a customer's child. If trust is betrayed and the child is left hungry and crying for several hours, it is probable that the services of the plumber will not be welcome in the future, notwithstanding the quality of the person's work as a plumber. In the same manner, even when the entrepreneur is confident that the system functions in principle, not all participants might be trusted, inducing the person to withdraw from the network activities. In certain respects *every social relation within the network is simultaneously both systemic and personal.* Thus, frames of both confidence and interpersonal trust are tacitly present all the time in social interactions.

The structure of the network also influences the development of confidence and trust within the network. Interdependencies and power relations in the network are not necessarily symmetrical; an actor holding a central position in the network may pose more credible sanctions than those in inferior positions. These asymmetrical interdependencies are some of those accepted vulnerabilities covered by the sphere of trust. As demonstrated above, seeking flexibility and efficiency through networking can either be based on horizontal ties between equal actors or on vertical patron-client relationships, which are common in subcontracting. There are more 'structural holes' in vertical networks. According to Ronald Burt's (1992) work on the issue, the holder of the central position in the network may wield strategic influence over others – control the flow of information and collect extra profit. Thus, for example, Robert Putnam puts more emphasis on the importance of horizontal networks in building trust. According to him, a vertical network cannot sustain social trust and co-operation because of asymmetrically distributed power (Putnam *et al.,* 1993, pp.173-174).

Industrial Networking in Rural Finland

As noted above, industrial networking has been promoted as a sustainable business strategy for small rural firms in Finland. If we take a closer look at the model of networking recommended by most authorities on Finnish rural policy, we recognize that it means, basically, vertical subcontracting. In this model the leading companies outsource the inferior tasks to small enterprises in rural areas where employment opportunities are scarce. At the same time, they take control of the networks since they command the key functions – design, product development and marketing – and control the central positions and the flow of information in the network structure (Ruuskanen, 1999).

The progress of this model has been slow, however. Although Finnish companies are forerunners in Europe in the spread of team working and in the use of information technology, partnership-type production cooperation among Finnish firms is still relatively rare (Alasoini, 2000). In Finland networking is more common between large firms: among the smallest firms, in particular, there is a great deal of suspicion towards networking. Instead of networking, small-scale enterprises favour business activities in which they can control production and marketing by themselves (LTT, 1995; CRP, 1996, p.109). Attitudes most common among the interviewees can be summarized as follows: *it is true that networking and collaboration are good ideas in principle, but they just don't fit to my own business activities.* What, then, leads to this attitude, which acknowledges the benefits of co-operation in principle, but at the same time is overtly negative towards the entrepreneurs' personal engagement in co-operative networks?

Individualization and Eroding Confidence

'Independence', 'freedom' and 'autonomy' are all concepts widely connected with the values and the motives of small-scale enterprising (Kankaanpää and Leimu, 1982; Scase and Goffee, 1982; Curran, 1986; Vesala, 1992). The ideal of autonomy and independence is clearly evident also in my data. These motives become manifest particularly in situations where the entrepreneurs compare their life and activities with their previous experiences as wage-workers. For them, wage-work signifies social restrictions – routine tasks, hierarchies between supervisors and workers, and strict working hours. The idea of being one's own master, independent of timetables and controls imposed by supervision, are central elements in the identity of rural entrepreneurs. They would like to decide 'who the customers are, what the products are, and how the production is technically executed', as one interviewee expressed it.

Becoming a self-employed entrepreneur is buying one's own autonomy, and the right to manage one's own work. The price of the autonomy is, however, uncertainty and risk. Unlike a wage-worker, self-employed entrepreneurs are clearly on their own. They, alone, bear the economic and social risks of the business. Because of the nature of business, personal income can be irregular. The same holds true also for working hours. When there are orders and customers, entrepreneurs have to work long hours. During quiet seasons there is no work or income. Bearing these fluctuations by themselves may often cause anxiety, as illustrated below:

All you've got is your own money. I recall how many times I've been thinking [that] today we have sold five thousand, but how about tomorrow? What will happen the next month, if nobody buys anything and the phone won't ring? What will happen? How will we make our living then? There was always the anxiety and never satisfaction, the future was [just] a source of concern all the time.

The contradiction between the two basic elements of self-employment is evident in narratives such as the one above. The idea of autonomy is seen hand-in-hand with uncertainty and constraint. Despite a certain degree of autonomy, everyday work with its obligations is highly restrictive of the freedom of a self-employed entrepreneur. Besides performing the actual tasks of production or serving customers, self-employed entrepreneurs usually have to spend hours doing paperwork. To increase one's freedom, it might often be rational to expand one's business. Usually this involves hiring employees and borrowing capital for the necessary investments, which in turn increase the risks. As one interviewee explained:

If I had a well-motivated employee, it would give me more freedom. I wouldn't have to do everything by myself, but I could concentrate more on planning and leading [the company]...when I do everything by myself, I know that even if there will be a bad business year I only have to manage the bank loan and the heating of the building. That's it, that's all I need.

One of my interviewees was Keijo, who had employed himself in a carpenter's shop. Keijo told me, that during the two years he had run his own business, he had worked 10 hours a day, six days a week. Despite this, he had not drawn any personal income from his business. The problem was that the local demand for his products was limited and he didn't own a van, so he could not transport his products to Southern Finland, where he knew there would be demand for traditional period furniture. When I inquired why he did not purchase a van, he answered that he would – as soon as he has saved money for it. 'Why do not you just borrow the money?' I asked. His answer is illustrative:

It might be profitable, yes...but...I don't know, my ambitions are so low. I prefer to act cautiously and slowly. It is so much easier to live without debts.

Rural entrepreneurs frequently viewed the idea of expanding their business oppressive. Even small debts that might be rational from the viewpoint of the business were regarded as hard to bear. Apparently the entrepreneurs wanted to avoid risk and dependency caused by loans, and by paid labour, at all cost. They would rather finance investments from their own income and reduce their own consumption. And the reason is clear: if times get harder, it is easier to manage, if there are no external obligations, such as employees and debts. Consequently, their basic lack of confidence forces them to keep their business small in scale.

Trust, Confidence and Collaboration

In very small enterprises, entrepreneurs bear practically all the risks by themselves. In that sense, they are the manifestation of individualized risk society. The positive dimension of individualization is the increased degree of freedom to act independently. At the same time, being on one's own – the individualization of risks and opportunities – may start eroding one's confidence. My interviewees did not even want to expand their business, but try to get along on their own devices. They work long hours and scrape a living. For these people getting along on one's own signifies autonomy, independence and freedom – the mythical connotations of entrepreneurship.

The eroding of confidence, and desire for independence, also influence their attitudes towards networking. As Vesala (1996, p.158) has noted, collaboration does not necessary threaten the autonomy of entrepreneurs: in fact it might support them in the pursuit of autonomy. If successful collaboration brings the firm regular customers or expands the circle of new customers, it helps to reduce uncertainty and strengthens the firm's continuity and the entrepreneur's autonomy.

The experiences of networking among my informants are somewhat ambivalent. Interest is shown toward co-operation, because it might help balance the workload and income. One possible strategy is vertical networking – subcontracting by a core group of companies to small rural firms. When collaboration is based on the vertical networks, larger firms usually make use of subcontracting to provide more flexibility. There is a big difference, however, between the flexibility of 'active versatility' and that of 'passive pliability'. In the first type of flexibility small firms utilize their joint capacity to react and adapt to temporary market niches. The second type is characterized by an ability and readiness to submit to outside pressure and to accept long-term risks and lasting cutbacks in labour standards and profits (Semlinger, 1993, pp.164-168).

If we examine the networking practices experienced by rural entrepreneurs, we recognize that the latter type of flexibility is familiar to them. Delivery time is usually very short. Even established agreements require an ability to deliver large orders promptly. When the 'lead company' makes an extra order, small suppliers have to adapt their functions to the demand. The need for pliability tends to trickle down in the vertical networks. Small-scale enterprises, which usually form the most peripheral part of the network, have to adapt to the requirements of the larger firms and fill the orders, even if it means working at night and on the weekend. In the vertical networks, small rural suppliers, furthermore, often produce simple components, manufacturing of which is routinized semi-skilled work. It seldom requires specialized knowledge, since leading companies design the components. This does not fit the idea of being one's own master, independent from strict timetables and control imposed by supervision, as described below:

> I keep myself busy with the new projects that make the work meaningful. I like to make experiments with new things, new materials. I get sick if I have to do mass production all the time, you don't have to think about it or plan anything, you just do it like some kind of machine.

As a consequence, entrepreneurs have contradictory attitudes toward subcontracting. They do supply components to larger companies, but at the same time they try to avoid dependency on larger firms. Instead, they often attempt to design their own finished products to be able to put an end to subcontracting.

Discussion

It is often stated that the main problem in building industrial networks is the lack of trust. In this chapter I have suggested that in rural areas it is not the lack of trust but confidence that prevents entrepreneurs from engaging themselves in networks and joint enterprises. Apparently vertical networking – the process of concentration without centralization (Harrison, 1997) – does not sustain confidence. Rural entrepreneurs are reluctant to participate in networks, because they feel their structural position in the network is subordinate and insecure. Due to the lack of confidence, entrepreneurs avoid taking risks. They do not expand their businesses, but work long hours by themselves and rely on their family members. The insecure position at the margins of the economy does not leave much room for developing trust either. Their motives are often aesthetic, such as freedom, independence, and autonomy. As a consequence, they withdraw from joint activities.

I suggested earlier that both trust and confidence are present when entrepreneurs make decisions concerning their activities. By reflecting on the above treatment of business practices of rural entrepreneurs, we might array the attitudes toward co-operation in a two dimensional taxonomy, as in Figure 15.1.

TRUST

	High	I. The Networker	II. The Classical Entrepreneur
C		Committing to common activities	Individual's need for achievement
O			Opportunism in co-operation
N		Sharing risks among the	
F		members of the	
I		networks	
D			
E	Low	III. The Small Scale Enterpriser	IV. The Self-employed
N			Uncertainty and anxiety
C		Avoiding risks	
E			Withdrawal from common
		Small scale networking	activities

Figure 15.1   Categories of Trust and Confidence

Along the vertical dimension there is the continuum of confidence, which depends on one's position in any known abstract system, such as an industrial network. If the position in the network is subordinate or marginal, the entrepreneur feels his or her future is uncertain and therefore lacks confidence. If the entrepreneur is positioned in the core of the system he or she feels more stable and confident. Along the horizontal dimension, I have displayed the continuum of trust (the extent an actor trusts other people). There are four typical categories:[2]

- In the first group we find a networker with a high level of confidence and a high level of trust. The networker shares risk with partners and dares to commit to networks;
- The classical 'Schumpeterian' entrepreneur is driven, basically, by self-confidence and the need to achieve. The classical entrepreneur lacks trust, but may participate in networks since he or she is confident that problems can be prevented with credible sanctions;
- The small-scale enterpriser lacks confidence and is therefore reluctant to expand his or her business. This type is willing to co-operate, but at the same time finds it difficult, since it would usually require investments and changes in production; and
- In the fourth quarter, the self-employed entrepreneur lacks both confidence and trust, which causes uncertainty and anxiety. The person wants to minimize risks and the goal is simply to earn a living by doing what that person likes doing. Often such people entertain thoughts of giving up, but the idea of autonomy usually keeps them going.

The globalizing economy is more-and-more characterized by its web-like industrial structure. In such a structure, as Castells (1996, p.96) puts it, the most important element for successful managerial strategy is to position a firm in the web in such a way as to gain competitive advantage for its relative position. The Networker, with high trust and high confidence, seems to be an ideal actor in the global network structure. Piore and Sabel were possibly thinking of him/her when they outlined the model of flexible specialization. There seems to be, however, mainly marginal network positions left for small-scale rural enterprises. At the margins of the network economy, in remote rural areas, especially, the erosion of the entrepreneurs' confidence and trust is quite evident. The identity of small-scale businessmen as independent entrepreneurs, together with the erosion of confidence and trust, seems to prevent them from commitment to networks and marginalizes them further from the global network economy.

Conclusion

It is well known that both cultural change and the building of trust are long and difficult processes. If rural policy seeks to revitalize the countryside by encouraging industrial networking it might be aimed at supporting the confidence of rural entrepreneurs. This could be achieved by supporting their long term market position in the network economy – that is, by encouraging their own innovations, technologies, labour and marketing skills, not by supporting vertical networks that centralize power and know-how in the leading companies of the networks.

Notes

1   The interviewees were engaged in woodworking and food processing industries – the branches emphasized in Finnish rural policy strategies. A total of 23 interviews were conducted. All the interviewees had been approached by a representative of various public projects encouraging networking and many of them had prior experiences of co-operation.
2   The categories are Weberian 'pure types' and they are not to be found in real life as such. None of my interviewees belong unambiguously to any of these categories.

References

Alasoini, T. (2000), 'Matkalla Kohti Oppivia Organisaatioita? Tiimityö ja Tuotannollinen Verkostoituminen Suomessa Tietoyhteiskunnan Kynnyksellä', *Hallinnon tutkimus,* 1/2000, pp. 6-18.

Atkinson, J. (1987), 'Flexibility or Fragmentatation? The United Kingdom Labour Market in the Eighties', *Labour and Society,* vol. 12(1), pp. 87-105.

Baier, A. (1986), 'Trust and Antitrust', *Ethics,* vol. 96, pp. 231-260.

Barbalet, J. (1998), *Emotion, Social Theory, and Social Structure: A Macrosociological Approach,* Cambridge University Press, Cambridge.

Beck, U. (1993), *Risk Society: Towards a New Modernity,* Sage, London.

Beck, U. (1994), 'The Reinvention of Politics: Towards a Theory of Reflexive Modernization', in U. Beck, A. Giddens and S. Lash (eds), *Reflexive Modernization: Politics, Tradition and Aesthetics in the Modern Social Order,* Polity Press, Oxford, pp. 1-55.

Burt, R. (1992), *Structural Holes: The Social Structure of Competition,* Harvard University Press, Cambridge.

Castells, M. (1996), 'The Rise of the Network Society. The Information Age', *Economy, Society and Culture,* vol. 1, Blackwell, Cambridge.

Coleman, J. (1988), 'Social Capital in Creation of Human Capital', *American Journal of Sociology,* vol. 94, pp. 95-120.

Co-Operative Group of Rural Policy (CRP)(1996), 'Maaseutuohjelma. Toimiva maaseutu. Maa- ja Metsätalousministeriön Asettaman Työryhmän Ehdotukset ja Perustelut', *Maaseutupolitiikan yhteistyöryhmän julkaisu,* vol. 1/1996.

Curran, J. (1986), *Bolton Fifteen Years On: A Review and Analysis of Small Business Research in Britain 1971-1986,* Small Business Research Trust, London.

Dasgupta, P. (1988), 'Trust as a Commodity', in D. Gambetta (ed.), *Trust: Making and Breaking Cooperative Relations,* Basil Blackwell, Oxford.

Giddens, A. (1991), *Modernity and Self-identity: Self and Society in the Late Modern Age,* Polity Press, Cambridge.

Granovetter, M. (1985), 'Economic Action and Social Structure: The Problem of Embeddedness', *American Journal of Sociology*, vol. 91, pp. 481-510.

Granovetter, M. (1992), 'Economic Institutions as Social Constructions: A Framework for Analysis', *Acta Sociologica*, vol. 35, pp. 3-11.

Hakim, C. (1990), 'Core and Periphery in Employers' Workforce Strategies', *Work, Employment and Society*, vol. 4(2), pp. 157-188.

Harrison, B. (1997), *Lean and Mean: The Changing Landscape of Corporate Power in the Age of Flexibility*, The Guilford Press, New York.

Kankaanpää, A. and Leimu, H. (1982), 'Yrittäjien Käsitykset Yritystensä Perustamissyistä ja Merkityksestä Pienteollisuudessa', *Turun Kauppakorkeakoulun Julkaisuja*, A-8, Turku.

Lewis, D. and Weigert, A. (1985), 'Social Atomism, Holism, and Trust', *Sociological Quarterly*, vol. 26, pp. 455-471.

Lorenz, E. (1988), 'Neither Friends nor Strangers: Informal Networks of Subcontracting in French Industry', in D. Gambetta (ed.), *Trust: Making and Breaking Cooperative Relations*, Basil Blackwell, Oxford, pp. 194-210.

LTT (1995), 'Yrittäjien Yhteistyöasenteet ja Verkostoituminen', *Liiketaloudellinen Tutkimuslaitos*, B 118, Helsinki.

Luhmann, N. (1979), *Trust and Power*, John Wiley and Sons, New York.

Luhmann, N. (1988), 'Familiarity, Confidence, Trust: Problems and Alternatives', in D. Gambetta (ed.), *Trust: Making and Breaking Cooperative Relations*, Basil Blackwell, Oxford, pp. 94-107.

Niittykangas, H. (1992), *Maaseudun Yritystoiminnan Kehittymismahdollisuudet*, Jyväskylän Yliopisto, Jyväskylä.

Piore, M. and Sabel, C. (1984), *The Second Industrial Divide: Possibilities for Prosperity*, Basic Books, New York.

Powell, W. (1996), 'Trust-based Forms of Governance', in R. Kramer and T. Tyler (eds), *Trust in Organizations*, Oxford University Press, Oxford, pp. 51-67.

Putnam, R., Leonardi R. and Nanetti, R. (1993), *Making Democracy Work: Civic Traditions in Modern Italy*, Princeton University Press, Princeton.

Ruuskanen, P. (1999), 'Rural Development Policy in Finland in the 1990s: Toward Flexible Specialization or Spatial Taylorism?', in C. Kasimis, and A. Papadopoulos (eds), *Local Responses to Global Integration*, Ashgate Publishers, Aldershot, pp. 223-244.

Sabel, C. (1989), 'Flexible Specialisation and the Re-emergence of Regional Economies', in P. Hirst and J. Zeitlin (eds), *Reversing Industrial Decline?: Industrial Structure and Policy in Britain and Her Competitors*, Oxford, Berg, pp. 17-70.

Sako, M. (1992), *Prices, Quality and Trust: Inter-firm Relations in Britain and Japan*, Cambridge University Press, Cambridge.

Sako, M. (1998), 'Does Trust Improve Business Performance?', in C. Lane and R. Bachmann (eds), *Trust in and Between Organizations*, Oxford University Press, Guildford, pp. 88-117.

Scase, R. and Goffee, R. (1982), *The Entrepreneurial Middle Class*, Croom Helm, London.

Seligman, A. (1997), *The Problem of Trust*, Princeton University Press, Princeton.

Seligman, A. (1998), 'Trust and Sociability: On the Limits of Confidence and Role Expectations', *American Journal of Economics and Sociology*, vol. 57, pp. 391-314.

Semlinger, K. (1993), 'Small Firms and Outsourcing as Flexibility Reservoirs of Large Firms,'in G. Grabher (ed.), *The Embedded Firm: On the Socioeconomics of Industrial Networks*, Routledge, London, pp. 161-178.

Vesala, K. (1992), 'Pienyrittäjien Kontrollipremissi: Sosiaalipsykologinen Tarkastelu', *Acta Psychologica Fennica, Soveltavan Psykologian Monografioita*, 5.

Vesala, K. (1996), 'Yrittäjyys ja Individualismi: Relationistinen Linjaus', *Helsingin Yliopiston Sosiaalipsykologian Laitoksen Tutkimuksia*, 2/1996, Helsinki.

Woolcock, M. (1998), 'Social Capital and Economic Development: Toward a Theoretical Synthesis and Policy Framework', *Theory and Society*, vol. 27(2), pp. 151-208.

Index

actor networks 182-184
agribusiness 11, 17, 43-55, 70, 103-117, 119-133, 153-167, 235
agricultural restructuring 16, 173-187, 189-200, 225-232
agriculture 11, 18, 57, 89-98, 206, 221-237
    alternative 69
    organic 15, 19, 69, 70, 81-82, 141, 149
agrifood
    networks 16, 18, 57-67
    politics 59, 149
    systems *see* agribusiness
agro-industrialization 69, 119-133, 140
anarchy 7
Argentina 18, 119-133
Australia 12, 21, 89-98, 221-237

bananas 60, 63-65, 70
beef 17, 90
biodiversity 11, 12, 37-39, 70, 79
bioregionalism 15
biotechnology 48, 74-77, 129-130
    see also food, genetically modified
Brazil 15, 18, 43-55, 135-150

chickens 103-117
Canada 57, 60, 62, 72, 81, 138, 182
Chile 11
China 7, 81
class 7, 90
coffee 60-65
communities
    coastal 173-187
    mining *see* mining
    mountain 173-187
community
    development 21, 208, 209, 221-237
    resistance 6, 19, 103-117, 176, 244
    wellbeing 116, 182, 225
consumers 10, 43, 61, 136, 140
consumption 14, 16, 25-40, 62

contract farming 126
co-operatives 19, 22, 63, 140, 153-167, 176, 180
corporations *see* TNCs
culture 36-40

democracy 15
dependency 7, 80, 206
deregulation 6, 17, 138, 153, 156-158, 167, 173, 190, 205
'developmentalism' 69, 71, 79, 83, 137

eco-consciousness 178
ecology *see* environment
economic rationalism *see* neoliberalism
education 8, 37, 139, 146, 149, 206, 217, 223-232, 23-237
employment 8, 165-167, 213, 214, 233, 234, 249
entrepreneurs 21, 22, 241-251
environment 6, 8, 11, 17, 61, 69, 71, 81, 83, 96, 97, 148, 182, 183, 207, 208
environmental
    degradation 6, 11, 12, 17
    pollution 14, 106, 108-110
    protection 13
Europe 20, 27, 62, 70, 73, 81, 205
European Union 4, 25, 48, 62
exclusion
    see social exclusion

fair trade 14, 16, 57-67, 70
farming *see* agriculture
finance 20, 22, 91, 94-96, 190-193, 201, 206
Finland 21, 241-254
fiscal crisis of the state 195
fishing 178-187
food 11
    differentiation 15, 25-40, 40
    fast food 25-40, 128
    food chain 17, 43-55

genetically modified 17, 48, 59, 70, 74, 75, 81-82, 84, 130
labelling 17, 59
packaging 61
processing 43-55, 119-133
quality 138, 139, 141, 156
safety *see* food security
security 17, 69-84
slow food 14, 15, 25-40
standardization 15, 25-40
Fordism-postfordism 103-117
France 25, 138, 155
fruit and vegetables 72, 73, 120, 122, 128, 198

(GATT) General Agreement on Tariffs and Trade 72, 89
genetics *see* food, 59
gentrification 207
global
    circuits 7, 197
    events 9
    finance 20
    investment 7
    regulation 58, 70, 75, 84, 89-98, 158, 159, 167, 196
    sourcing 78, 103
    trade 57-67, 80, 91, 153-167, 190
globalization 3-22, 58, 70, 71, 113, 114, 119-133, 153-167, 173-187, 205-218, 221-237, 244, 253
    coping strategies 173-187
    definition 3, 182, 221
    'globalization project' 5, 114
global-local interaction 3-22, 54, 57-67, 70, 83, 117, 131, 173-187, 189-202, 206, 221-237
glocalization 10
governance *see* regulation
Great Britain *see* United Kingdom
'greening' 20, 82

health 10, 15, 38, 61, 74, 107, 109, 111, 206, 225-233
high fructose corn syrup *see* sugar
housing 214-216

income 7
India 7, 13, 25
Indigenous peoples 90
Indonesia 11

inequality *see* social inequality
information technology 45, 46, 235
investment 61
Italy 31, 33, 35

Japan 4, 57, 60, 62, 138

Korea 4

labour issues *see* workers
Latin America 7, 138
leadership 234, 235
Less-Favoured Zones 189-202
life chances 7
local-global *see* global-local
localization 36-40, 221, 222, 223

McDonaldization 9, 25-40
media 9, 108, 179
Mexico 73, 78
milk 18, 46, 120, 128, 129, 135-150
mining 19, 173-187
modernization 10

neoliberalism 5, 14, 17, 58, 89-98, 132, 205
network economy 241-254
New Zealand 6, 7, 17, 89-98
North/South relations 16, 57-67, 71, 72, 76, 77, 78
(NAFTA) North American Free Trade Association 4, 25, 48, 72, 78
Norway 19, 173-187

organic *see* agriculture

policy *see* rural policy
politics 208, 209
    see also agrifood politics
postmodernism 26
potatoes 128
poverty 209-213
power relations 69-84, 210, 248
privatization 6, 193, 205, 236
protest
    see social protest

rationalisation 31, 32, 173
regulation *see* global regulation
resistance *see* community resistance

restructuring *see* agricultural
    restructuring
risk 222, 236, 244
rural
    policy 17, 21, 94, 205-218, 236, 241-
    254
    restructuring 189-202, 225-32

social
    capital 14, 21, 182-184, 200, 207,
    245
    exclusion 7, 14, 20, 147, 198, 205-
    218, 236
    impacts 135-150
    inequality 14, 58, 83, 139, 205-218
    resistance *see* community resistance
South Africa 19, 153-167
standardization 9, 10, 25-40
state 3-5, 20, 77, 78, 89-97, 103, 138,
    222, 224
    the role of 142-144, 189-202
    welfare state 6, 208, 209, 236
structural adjustment 80, 225, 227
sugar 15, 16, 43-55, 121, 122, 128
    high fructose corn syrup 47
supermarkets 62, 120, 120-126, 131,
    132, 155
sustainability 13, 14, 58, 74, 77, 79, 82,
    96, 97, 146, 149, 236
sustainable agriculture *see* sustainability
sustainable development
    see sustainability
sustainable livelihoods 3, 13-15, 115,
    116, 165, 166, 205-218
Switzerland 62

technological change 43-55, 119, 120,
    132, 135-150, 182-184
Texas 17, 103-117
tomatoes 73, 74
tourism 19, 233
trade *see* global trade
Transnational Corporations (TNCs) 4, 5,
    8, 16, 44, 47, 66, 72, 73, 78, 80, 103-
    117, 189-202
    see also agribusiness
'triple bottom line' 13
trust 21, 241-254

unemployment 132-133, 185, 192, 213,
    214
United Kingdom 81, 90, 92, 138, 155,
    205-218
United States 4, 17, 25, 29, 30, 39, 57,
    60, 61, 70, 72, 74, 82, 92, 103-117,
    138, 155

vegetables *see* fruit and vegetables

welfare state
    see state
wheat 17, 90
wine 19, 60, 121, 153-167
women 8, 21, 210, 212
wool 17, 90
workers 8, 61, 62, 71, 74, 75, 130, 131,
    140, 146, 147, 148, 155, 165-167,
    176, 205-207, 213, 214, 217, 218
(WHO) World Health Organization 11
(WTO) World Trade Organization 5, 8,
    70-76, 78 80, 196